Metabolic Engineering for Bioprocess Commercialization

Stephen Van Dien
Editor

Metabolic Engineering for Bioprocess Commercialization

Springer

Editor
Stephen Van Dien
Genomatica, Inc.
San Diego, CA
USA

ISBN 978-3-319-41964-0 ISBN 978-3-319-41966-4 (eBook)
DOI 10.1007/978-3-319-41966-4

Library of Congress Control Number: 2016943877

Printed on acid-free paper

This Springer imprint is published by Springer Nature
The registered company is Springer International Publishing AG Switzerland

Contents

Chapter 1
Introduction

Stephen Van Dien

Abstract The field of metabolic engineering, loosely defined as the manipulation of living organisms to achieve a desired metabolic objective, has grown and advanced significantly over the past 20 years. First applied to improve organisms producing existing biochemicals, it is now a promising approach to develop biocatalysts for the production of nonnatural fuels and chemicals previously accessible only through petrochemical processes. New tools such as gene synthesis, advanced cloning techniques, 'omics' analysis, and mathematical modeling have greatly accelerated the pace of innovation in the field, leading to many success stories and even some commercialization examples. This volume reviews the current state of the art in tools and technologies for metabolic engineering.

In October 1996, a group of 100 or so microbiologists, molecular biologists, and chemical engineers gathered at a conference resort outside of Boston to discuss the emerging field of Metabolic Engineering. Recombinant DNA technology was well established at this time, particularly for the expression of industrial and pharmaceutical proteins, and the idea of manipulating an organism's metabolic network was starting to take shape. As defined in a landmark paper by the late Professor James 'Jay' Bailey several years earlier (Bailey 1991), metabolic engineering was distinguished from genetic engineering by the need to express multiple genes that form a pathway, rather than a single protein. The intended product is not the enzymes themselves, but rather the result of their function; i.e., metabolism. With the belief that all metabolic pathways had been elucidated already, metabolism research had fallen out of fashion in the 1970s and 1980s, while structural biology and genetic regulation became the hot topics. Even the undergraduate biochemistry course I had taken glossed over the chapters on amino acid and nucleotide biosynthetic pathways, to spend more time on molecular biology. Technology for manufacture of biochemicals, from ethanol to amino acids to antibiotics, was largely mature. Production organisms were obtained using classical mutagenesis and

S. Van Dien (✉)
Genomatica, Inc., 4757 Nexus Center Drive, San Diego, CA 92121, USA
e-mail: svandien@genomatica.com

S. Van Dien (ed.), *Metabolic Engineering for Bioprocess Commercialization*,
DOI 10.1007/978-3-319-41966-4_1

screening, starting from a native producer, and most of the development in biochemical engineering was on cost optimization of the process, not the organism. Many of the early pioneers in metabolic engineering research were trained in fermentation, process design, and separations. When directed genetic manipulations became commonplace in the late 1980s for the development of modified enzymes via site directed mutagenesis, it wasn't long before the implications to modifying entire pathways were realized. There was a resurgence of interest in metabolism, and a new field was born.

I had the privilege of attending this conference in Massachusetts, now known as Metabolic Engineering I, as a graduate student; I was nearly star-struck by the opportunity to not only see lectures by, but also to share casual dinner conversation with top researchers in my field. Before the meeting, I had a vague idea that my thesis project to selectively control the expression of two genes, encoding the synthesis and degradation of a simple polymer, was considered metabolic engineering. I knew very little about the implications pathway manipulation could have for biochemical production. Being from Texas, and with an undergraduate degree in Chemical Engineering, my naïve understanding was that all chemicals came from oil and gas. By the end of the meeting, I had found a scientific home. Not only was I fascinated with the unlimited potential of microbial metabolism, but I also loved the intersection of mathematical and engineering concepts with molecular biology. A cartoon shown in one of the presentations joked that the difference between genetic engineering and metabolic engineering is "lots and lots of math".

Although the science has advanced significantly in the last 20 years, some of the themes of that first meeting are still relevant today. A major area of focus was the development of computational tools for both measurement and design. Just as mechanical or chemical engineers use theory and mathematical models to design products and processes, metabolic engineers sought to follow the same paradigm. Flux balance analysis (Schilling et al. 1999; Schilling et al. 2000) and metabolic control analysis (Fell and Sauro 1985) had engineering parallels in process optimization and process control, respectively. The underlying theory, based on the stoichiometry of metabolic networks, had been developed more than 20 years earlier by Kacser and Burns (1973), and was now finding application not only in understanding metabolism, but also in manipulating metabolism. Still relevant today, stoichiometric modeling has been used to guide development of novel organisms for commercial bioprocessing (Yim et al. 2011). Introducing an exogenous pathway, or enhancing expression of a native one, is not sufficient for achieving commercial metrics. Metabolic engineering involves re-routing central metabolism to supply precursor molecules, balancing reducing equivalents, and proper tuning of the pathway expression. Once basic strain designs are generated with modeling, genetic tools are needed to implement designs and experimental techniques required to measure the output. Artificial promoters, gene integration techniques, transcript profiling, and the elucidation of complex regulatory circuits were all discussed at the 1996 meeting. Applications of metabolic engineering covered a broad spectrum. Key products of interest were pharmaceuticals, amino acids, modified fatty acids, the natural biodegradable polymer polyhydroxybutyrate, and solvents. Perhaps envisioning the cellulosic biofuels boom that would come a decade later, work was presented on C5/C6 sugar co-utilization. Notably absent were nonnatural chemicals such as isobutanol or 1,4-butanediol. Other applications of metabolic engineering principles included the study of human tissues, cell culture, plants, and xenobiotic degradation. Nowadays, with a few exceptions the term metabolic engineering is usually understood to mean applications in microbial bioproduction.

With the turn of the century came the era of functional genomics, where nearly complete genome sequences and microarrays of model organisms were commonplace. Gene sequencing (Sanger) and oligonucleotide synthesis were available as services, many molecular biology protocols could be purchased as kits, and PCR machine throughput increased; consequently, the pace of research accelerated. Success stories like 1,3-propanediol (Nakamura and Whited 2003) were starting to emerge, catching the attention of researchers and chemical companies alike. The excitement and opportunity around metabolic engineering was captured by Professor Bailey in a video presentation of his acceptance speech for the First Merck Award in Metabolic Engineering at the Metabolic Engineering III conference in October, 2000, a few months before his death (Bailey 2001). His vision was captured in a song:

The (Metabolic Engineering) Times, They Are A 'Changin'
(Bob Dylan/Jay Bailey)

* Reproduced from Stephanopoulos (2001)

Come gather Metabolic Engineers 'cross the land
At MEIII we'll take command
Of cells that are too slow to produce or grow.
If it's higher fluxes you're needin'
Then we'll shift the controls, and block bad outflows.
For the times, they are a changin'.

Do you need a new molecule or neutraceutical
The Metabolic Engineer has the answers for you.
We'll import new pathways, and shuffle them too.
Is your lead compound library fadin'?
We'll give new adducts to your old natural products
For the times, they are a changin'.

Rational or random, which way is best?
Solving the problem passes the test.
Complex responses confuse the quest.
More genetic and array technologies
Will give us insights to networks' delights.
For the times, they are a changin'.

Genomes are in hand, the sequences there,
An amazing resource that we all share.
Genes and controllers, bioinformatics tells us where.
But how is all of this workin'?

Let's decipher a yeast, understand that at least.
For the times, they are a changin'.

How is phenotype controlled by the genes?
Nobody knows, least of all the machines.
Medicine will thrive if we can discover the means,
To merge our knowledge and information
And find genes' intent and control by environment.
For the times, they are a changin'.

Metabolic Engineers have all the tools
Biology, computing, and engineering rules,
Knowledge, experience, perspective on detail.
Let's help Metabolic Genomics to set sail.
Opportunity's here . . . but now it's time for a beer!
For the times, they are a changin'.

The times continued to change. The next decade brought NextGen sequencing, gene synthesis, and the rise of synthetic biology. The number of genomes sequenced, and consequently the unique gene sequences in GenBank and other databases, grew exponentially. This provided more choices for assembling genes into pathways, and these genes could now be custom ordered with codon-optimized sequences to improve expression in the host production organism. Inexpensive whole-genome sequencing enabled the deconvolution of mutations from adaptive evolution experiments, as well as RNAseq to routinely monitor transcriptome profiles even more efficiently than microarrays. Other 'omics' techniques—proteomics, metabolomics, and ^{13}C-flux analysis—also found applications for diagnosing bottlenecks in production strains (Van Dien 2013). New success stories for diverse products including artemisinin (Westfall et al. 2012), 1,4-butanediol (Yim et al. 2011), lactic acid, and isobutanol (Atsumi et al. 2009) emerged.

Changes in our approach to metabolic engineering are still on the horizon, driven both by technological advances that improve the way we do science and market forces that influence the product choices. This volume includes workflows and methodologies based on state-of-the-art technology in 2016. There will certainly be cases where the methodologies described are no longer practiced several years from now; although the specific methods may change, the overall scientific approaches to engineering organisms will continue to be relevant. Specifically, the articles that follow describe approaches to pathway development and tuning, host strain optimization, evolution to overcome stresses, and fermentation process scale-up. Theory and modeling are not covered here; the reader is referred to excellent textbooks on the subject (Stephanopoulos et al. 1998; Patnaik 2013) and the references therein. By application of the methods described in the following sections, we hope to inspire a new generation of metabolic engineering success stories.

References

Atsumi S, Li Z, Liao JC (2009) Acetolactate synthase from Bacillus subtilis serves as a 2-ketoisovalerate decarboxylase for isobutanol biosynthesis in Escherichia coli. Appl Environ Microbiol 75:6306–6311

Bailey JE (1991) Toward a science of metabolic engineering. Science 252:1668–1675

Bailey JE (2001) Reflections on the scope and the future of metabolic engineering and its connections to functional genomics and drug discovery. Metab Eng 3:111–114

Fell DA, Sauro HM (1985) Metabolic control and its analysis. Additional relationships between elasticities and control coefficients. Eur J Biochem 148:555–561

Kacser H, Burns JA (1973) The control of flux. Symp Soc Exp Biol 27:65–104

Nakamura CE, Whited GM (2003) Metabolic engineering for the microbial production of 1,3-propanediol. Curr Opin Biotechnol 14:454–459

Patnaik R (2013) Engineering complex phenotypes in industrial strains. John Wiley & Sons, Hoboken, NJ

Schilling CH, Edwards JS, Palsson BO (1999) Toward metabolic phenomics: analysis of genomic data using flux balances. Biotechnol Prog 15:288–295

Schilling CH, Edwards JS, Letscher D, Palsson BO (2000) Combining pathway analysis with flux balance analysis for the comprehensive study of metabolic systems. Biotechnol Bioeng 71:286–306

Stephanopoulos G (2001) In memorium: Jay Bailey (1944–2001). Metab Eng 3:393–394

Stephanopoulos GN, Aristidou AA, Nielsen J (1998) Metabolic engineering: principles and methodologies. Academic Press, San Diego, CA

Van Dien S (2013) From the first drop to the first truckload: commercialization of microbial processes for renewable chemicals. Curr Opin Biotechnol 24:1061–1068

Westfall PJ, Pitera DJ, Lenihan JR, Eng D, Woolard FX, Regentin R, Horning T, Tsuruta H, Melis DJ, Owens A et al (2012) Production of amorphadiene in yeast, and its conversion to dihydroartemisinic acid, precursor to the antimalarial agent artemisinin. Proc Natl Acad Sci U S A 109:E111–E118

Yim H, Haselbeck R, Niu W, Pujol-Baxley C, Burgard A, Boldt J, Khandurina J, Trawick JD, Osterhout RE, Stephen R et al (2011) Metabolic engineering of Escherichia coli for direct production of 1,4-butanediol. Nat Chem Biol 7:445–452

Chapter 2
Gene Expression Engineering

Nicholas J. Morse and Hal S. Alper

Abstract Cellular systems can become platforms for chemical production. Over the last four years, over 50 biopharmaceuticals have been approved for production, ranging in scope from hormones, enzymes, fusion proteins, antibodies, and vaccines. However, each of these applications—whether chemicals or pharmaceuticals, requires both a host organism and tools to engineer pathways in this chosen organism. The cellular hosts for these processes range in scope and complexity to include bacterial systems like *Escherichia coli*, yeast systems like *Saccharomyces cerevisiae*, and a variety of mammalian cell systems. To accomplish these production goals, it is necessary to control gene expression (especially of heterologous genes and pathways). This chapter will evaluate methods for controlling gene expression in the context of heterologous genes, endogenous genes, pathway expression and provide insight into new paradigms for flux control through gene expression circuits. A focus of this chapter will be on the various synthetic tools available for gene expression control. Although these basic principles are broadly applicable to multiple organisms, the predominant focus of this chapter will be on microbial systems, particularly *E. coli* and *S. cerevisiae*.

2.1 Introduction

Cellular systems can become platforms for chemical production. In 2004 and 2007, the U.S. Department of Energy and Pacific Northwest National Laboratory defined a set of top value-added chemicals which could be derived from biomass (Werpy et al. 2004; Holladay et al. 2007). These molecules can be used as precursors for a

N.J. Morse · H.S. Alper (✉)
Department of Chemical Engineering, The University of Texas at Austin,
200 E Dean Keeton St. Stop C0400, Austin, TX 78712, USA
e-mail: halper@che.utexas.edu

H.S. Alper
Institute for Cellular and Molecular Biology, The University of Texas at Austin,
2500 Speedway Avenue, Austin, TX 78712, USA

S. Van Dien (ed.), *Metabolic Engineering for Bioprocess Commercialization*,
DOI 10.1007/978-3-319-41966-4_2

wide range of industries, including transportation, textiles, food supply, environment, housing, recreation, and health. At the same time, as of 2014, there are over 200 approved biopharmaceutical products produced in cellular systems, accounting for well over $100 billion in sales (Walsh 2014). Over the last four years, over 50 biopharmaceuticals have been approved for production, ranging in scope from hormones, enzymes, fusion proteins, antibodies, and vaccines. However, each of these applications—whether chemicals or pharmaceuticals, requires both a host organism and tools to engineer pathways in this chosen organism. The cellular hosts for these processes range in scope and complexity to include bacterial systems like *Escherichia coli*, yeast systems like *Saccharomyces cerevisiae*, and a variety of mammalian cell systems. To accomplish these production goals, it is necessary to control gene expression (especially of heterologous genes and pathways).

As a result, selecting the appropriate expression host, understanding how these cellular systems function, and optimizing production of these pharmaceuticals and chemicals have become a focus in the field of metabolic engineering. Of utmost importance is the ability to control gene expression. Thus, this chapter will evaluate methods for controlling gene expression in the context of heterologous genes, endogenous genes, pathway expression and provide insight into new paradigms for flux control through gene expression circuits. A focus of this chapter will be on the various synthetic tools available for gene expression control. Although these basic principles are broadly applicable to multiple organisms, the predominant focus of this chapter will be on microbial systems, particular *E. coli* and *S. cerevisiae*.

2.2 Selecting Host Organisms and the Need for Heterologous Expression

The choice of host organism is often a major and first deciding factor in process optimization. Successfully importing heterologous pathways into organisms requires a delicate balance between the pathway of interest, the required pathway precursor availability, and the capacity for strong overexpression of foreign DNA. Bacterial systems, for example, have the ability to overexpress high quantities of proteins and enzymes (Liu et al. 2013a; Makrides 1996); however more complex biopharmaceuticals, which require more sophisticated posttranslational modification, are often expressed in higher eukaryotic systems (Walsh 2010, 2014; Sanchez and Demain 2012). Nevertheless, a wide range of cells are currently being explored for these broad applications, including bacterial cells, yeast cells, mammalian cells, insect cells, plant cells, and cell-free systems (Sanchez and Demain 2012). We highlight the three major classes of organisms chosen for metabolic engineering and chemical/protein production here.

First, the most popular choice for production especially in proof-of-concept experiments is the bacterium *E. coli*. The well-characterized genome and wide range of available synthetic tools make *E. coli* one of the most commonly used organisms for heterologous protein expression (Liu et al. 2013a; Sanchez and Demain 2012; Terpe 2006; Chou 2007; Rosano and Ceccarelli 2014). Rapid growth, high cell density fermentations, inexpensive media requirements, and a large range of available expression vectors make *E. coli* a prominent choice as a metabolic engineering host (Terpe 2006). For example, *E. coli* has been engineered to produce biofuel components like butanol and propanol, organic acids like lactic acid and succinic acid, amino acids like threonine and tryptophan, sugar alcohols like xylitol and mannitol, and a variety of drugs and polymers (Chen et al. 2013).

Yeast cells, on the other hand, have many of the advantageous of *E. coli* but provide the ability to perform eukaryotic posttranslational modifications. Much like their bacterial counterparts, yeast cells are economical, can reach high cell densities, and have been successful at producing high protein titers (Celik and Calik 2012). In addition, yeasts have the ability to perform posttranslational modifications like glycosylation, lack pyrogens, and viral inclusions which would be harmful in biopharmaceuticals, produce chaperonins that assist in protein folding, and have a higher tolerance to fermentative conditions (Sanchez and Demain 2012; Celik and Calik 2012; Liu et al. 2013b). As a result, yeasts are seeing an increase in use for the production of value-added chemicals. Their tolerance for acidic conditions makes yeast cells useful for the production of muconic acid, itaconic acid, and ricinoleic acid (Liu et al. 2013b; Blazeck et al. 2014a). In addition, they are seeing an increasing role in the large-scale biosynthesis of lipids and drug precursors (Ro et al. 2006; Blazeck et al. 2014b).

The last of the most commonly used expression systems for heterologous protein production are higher eukaryotic such as mammalian cells—especially for applications of biopharmaceutics as these cells have the capacity to secrete properly folded and glycosylated therapeutic proteins (Martinez et al. 2012; Nielsen 2013; Dalton and Barton 2014; Zhu 2012). Chinese hamster ovary (CHO) cells, for example, are among the most commonly used eukaryotic cells in biopharmaceuticals because of their ability to produce human-like proteins (Sanchez and Demain 2012; Martinez et al. 2012; Nielsen 2013). However, unlike their microbial counterparts, much less is known about their genetics, often limiting their uses to the expression of a single, protein product instead of an entire pathway.

Nevertheless, the wild-type cell rarely produces many of these valued metabolites at a high concentration. For example, *S. cerevisiae* cannot natively produce muconic acid from carbohydrates because of its lack of the necessary enzymes. However, by introducing and expressing the necessary genes for these enzymes inside the cell, the production of valuable precursor chemicals, like muconic acid, could be achieved in cellular expression systems (Curran et al. 2013a). As another example, the production of salvianic acid A in *E. coli* required both new enzymes to be introduced into the cell and certain native protein production to be stopped (Yao et al. 2013). In both cases, the cell may not produce the enzymes needed to convert intermediates in a pathway into the final product. Instead, these deficiencies were fulfilled by expressing heterologous proteins that can perform the needed function inside the host cell.

Thus, engineering entire pathways has two parts. One is the modification of the native cell machinery in order to force flux through a given pathway or prevent the degradation of an intermediate. The other is the expression heterologous proteins that are needed to perform the reactions the cell cannot natively catalyze. The next couple sections of this chapter will focus on the common approaches to performing both of these functions.

2.3 Modifying the Expression of Native Genes

Wild-type cells possess a cascade of regulatory elements that control the native transcriptome within the cell. However, most engineering efforts will force a cell to deviate from its typical, wild-type behavior, requiring genetic rewiring to do so. Thus, a major focus of gene expression control involves the modification of native gene expression. While the tools for both native and heterologous expression (described in the next sections) are similar, the rationale is quite different.

In contrast to heterologous expression, when modifying the expression of a native gene, it is important to realize that this gene, before any modifications, has some innate expression level inside the cell. The goal of engineering efforts is to modify (and potentially remove regulation) from this native expression pattern. If the relative amount of protein from the modified gene *increases* from the innate levels, it is said that the gene is overexpressed. Often times, this can be accomplished by either introducing multiple copies of the gene into the cell or modifying the regulatory region of a gene to swap in higher strength promoters and terminators (Nielsen 2013; Da Silva and Srikrishnan 2012; Redden et al. 2014). By this definition, nearly all heterologous protein expression is said to be a type of overexpression since the cell did not natively produce it prior to its introduction.

Alternatively, gene expression can be changed through a gene knockdown, which *decreases* gene expression relative to the innate levels. This can be accomplished by a total knockout of the gene, silencing the expression in the cell, or knocking down expression through methods such as RNA interference systems (Redden et al. 2014; Suess et al. 2012; Crook et al. 2014; Giaever and Nislow 2014) or promoter replacements. Essential genes that would otherwise inhibit growth of the cells (Giaever and Nislow 2014; Giaever et al. 2002; Winzeler et al. 1999) can be modified with knockdown techniques that can enable the elicitation of favorable phenotypes (Crook et al. 2014). Thus, unlike heterologous expression, a more delicate balance of expression is required when rewiring endogenous genes.

Yet, simply modifying the expression of native genes can be quite powerful and can lead to massive shifts in metabolic flux toward different products, thereby producing interesting or valuable products and phenotypes. For example, lipogenesis is an innate cellular process which converts intermediate products of sugar metabolism into fatty acids and triglycerides. By altering the expression of native enzymes, it is possible to rewire the flux through this pathway to make cells producing upwards of 90 % lipid content (Blazeck et al. 2014b).

Therefore, there is a need for altering gene expression—whether to create heterologous function or to enable rewiring of native function. Either way, a suite of synthetic tools is required to accomplish these goals. Thus, the remaining sections of this chapter will first cover how to introduce a new gene into the cell through either a plasmid-based expression system or a genomic integration. Next, strategies for increasing or decreasing the gene expression through approaches such as promoter engineering will be addressed. Following, we will look at regulation at the translational level in order to control the net protein production from expression. Finally, we will address emerging, sophisticated methods for flux control through gene expression control. Examples of applications to metabolic engineering will be addressed throughout.

2.4 Expression of Multiple Copies of a Gene Versus Higher Strength Promoters

DNA editing (either for heterologous pathways or other modifications) is typically performed in one of two ways: plasmid vectors or through genomic integrations. Plasmids are circular, double-stranded DNA molecules that replicate and express autonomously from the cell's chromosome. Therefore, they are useful in applications with only a handful of genes or as a tool for rapid prototyping and assessing the expression of a gene in a cell (Da Silva and Srikrishnan 2012; Madyagol et al. 2011). However, since multiple copies can be maintained in a cell at any one time, plasmid burden has been shown to affect the cell growth (Karim et al. 2013) and long-term stability is a challenge in bioprocessing. Genomic integrations, on the other hand, are typically used for applications requiring more stable and tight regulation on the expression of these genes (Da Silva and Srikrishnan 2012; Madyagol et al. 2011). Single or multiple copies can be inserted into the cell's genome at a time. Therefore, the relative strength of these genes will depend on the promoter and transcriptional regulation at their respective locus.

2.4.1 Plasmids

Plasmids are the most common, facile way to transfer genetic information into an organism (Berlec and Strukelj 2013). These elements are characterized based on their stability of replication, copy number in a cell, and segregation into daughter cells (Berlec and Strukelj 2013). These characteristics are controlled by the plasmid's replicon, promoters, selection markers, multiple cloning sites, and fusion protein tags (Rosano and Ceccarelli 2014). As such, a wide range of different combinations of these elements has been made, allowing for flexibility in the expression of a gene.

Autonomously replicating pieces of DNA have a replicon which contains both the origin of replication and the *cis*-acting control elements used to control the plasmid copy number in the cell (Rosano and Ceccarelli 2014; del Solar and Espinosa 2000). In *E. coli*, the three most commonly used origins of replications (ori) are the ColE1, the p15A origin, and the pSC101 origin (Rosano and Ceccarelli 2014). ColE1 is an origin of replication that has both a high-copy derivative and a low-copy derivative. The pUC plasmid series uses the high-copy (500–700 copies per cell) derivative of the origin of replication; the low-copy derivative (15–60 copies per cell) is used in the pMB1 plasmid series (Rosano and Ceccarelli 2014; Berlec and Strukelj 2013; Bolivar et al. 1976; Lee et al. 2006; Liang et al. 1999; Minton 1984; Sorensen and Mortensen 2005). However, since these plasmids use origins of replication from the same family, they compete for the same replicative machinery thus preventing maintenance of more than one unique plasmid (del Solar et al. 1998; Camps 2009). To create a multiple plasmid system, plasmids using origins of replications from different families are often used. For example, plasmids containing the p15A origin of replication can be used in combination with ColE1 plasmids when needing to express multiple plasmids in the same cell (Rosano and Ceccarelli 2014; Berlec and Strukelj 2013; Chang and Cohen 1978; Guzman et al. 1995; Nordstrom 2006). Additionally, other low-copy plasmids, such as the pSC101 series (≤5 copies per cell), are useful when the expressed protein is inhibitory or toxic to the cell (Stoker et al. 1982; Wang and Kushner 1991). Selection for cells maintaining the plasmids are then made through a variety of antibiotic gene markers. In *E. coli*, this is most readily done using antibiotic resistance genes such as ampicillin, chloramphenicol, kanamycin, or tetracycline (Rosano and Ceccarelli 2014; Berlec and Strukelj 2013). However, antibiotic-free plasmids have also been used to prevent the high cost of antibiotics at the industrial scale (Rosano and Ceccarelli 2014; Chen 2012; Goh and Good 2008; Hägg et al. 2004; Kroll et al. 2009, 2010, 2011; Peubez et al. 2010; Voss and Steinbuchel 2006; Zielenkiewicz and Cegłowski 2001).

Yeast plasmid copy number is also regulated by its origin of replication, much like *E. coli* plasmids. However, many of the common yeast plasmids act as shuttle vectors between *E. coli* and yeast, so they contain two origins of replication: one for yeast and one for *E. coli* (Da Silva and Srikrishnan 2012; Redden et al. 2014). Yeast origins of replications are usually either one of two types. The 2μ origin is often used for high-copy plasmids (≥10 copies per cell), while an autonomously replicating sequence and a centrometric sequence (ARS/CEN) are often used for low-copy plasmids (1–2 copies per cell) (Celik and Calik 2012; Da Silva and Srikrishnan 2012; Redden et al. 2014; Clarke and Carbon 1980). Yeast selection markers, unlike their *E. coli* counterparts, commonly use auxotrophic markers for selection. Yeast strains auxotrophic for leucine (*LEU2*), uracil (*URA3*), histidine (*HIS3*), lysine (*LYS2*), or tryptophan (*TRP1*) can carry auxotrophic markers on their plasmids as a selection for cells with the plasmid (Da Silva and Srikrishnan 2012; Redden et al. 2014). Drug resistance markers on vectors, such as resistance to Geneticin (*kanMX4*), nourseothricin (*natNT2*), and hygromycin (*hphNT1*), have also been used successfully for yeast plasmid systems for negative selections (Taxis and Knop 2006). However, much like *E. coli*, multiple plasmids containing the same origin of replication become burdensome to the cell (Futcher and Carbon 1986; Mead et al. 1986). As with *E. coli*, stability of these plasmids and the need for certain strains (e.g., industrial yeast strains that may not have the appropriate auxotrophy) are self-limiting when considering most industrial scale production needs.

One of the major challenges of a plasmid-based system is the metabolic burden of maintaining large copy numbers of plasmids inside the cell. The production of taxadiene, for example, saw that plasmid copy number for each expression module was a key determinant in the final titer (Ajikumar et al. 2010). The production of polyphosphate and lycopene demonstrate that high-copy plasmids caused such a burden on cells that low-copy plasmid alternatives actually increased overall concentration (Jones et al. 2000). And the biosynthesis of amorphadiene saw an improvement when modifying the plasmid copy number in the optimization of their pathway (Anthony et al. 2009). In each of these cases, plasmid copy number was attributed to a higher metabolic burden in the cell, leading to slower growth rates, lower cell density, and decreased processing. Instead, engineering a metabolic pathway requires a balance between gene expression and plasmid burden. Plasmid burden has been decreased by introducing expression cassettes into the chromosome of the host organism in a titrated fashion, and gene expression has been controlled through promoter engineering and translational optimization. This will be examined in the sections that follow.

2.4.2 Genomic Integration

The alternative to plasmid-based expression systems is genomic integration. While chromosomal modifications are considered more laborious compared to plasmids, genomic integrations offer greater stability over successive generations of growth and better control over expression. The most common approach to genomic editing is through the use of the cell's native DNA double-stranded breaks repair pathways, such as homologous recombination and nonhomologous end joining, followed by subsequent selection. Moreover, with the recent advent of CRISPR-Cas9 systems, targeted and efficient genomic integrations have become possible (Sander and Joung 2014).

Clustered Regularly Interspaced Short Palindromic Repeats (CRISPR) *Cas* proteins are a family of nucleases discovered in prokaryotes which specifically cleave DNA of a certain sequence. The most popularly used of these, protein 9 from the *Streptococcus pyogenes* CRISPR system (Cas9), is advantageous for its ability cause a double-stranded DNA based on the sequence of RNA bound to it, therefore promoting the above repair pathways (Sander and Joung 2014). Interchanging this RNA has allowed targeted integrations and knockouts in a wide range of hosts, including bacterial cells (Jiang et al. 2013), yeast cells (DiCarlo et al. 2013), and mammalian cells (Wu et al. 2014) via both homology direct and nonhomologous end joining pathways. As more *Cas* proteins are discovered, it is anticipated that this field of genomic manipulation will grow.

Other methods for genomic integrations, such as the use of recombinases and transposases, are also available for both bacterial and yeast hosts. Such methods promote the cleavage and rearrangement of DNA, often based on specific recognition sequences. In *E. coli*, the λ bacteriophage Red system or the prophage Rec system can introduce DNA flanked by short regions homologous to the host genome (Madyagol et al. 2011; Datsenko and Wanner 2000). The Cre-Lox system, shown to work in both prokaryotic and eukaryotic systems, induces recombination at a specific recognition sequence (Sauer 1987). Or the use of transposases, such as the Tn7 transposase in *E. coli* (Silva-Rocha and de Lorenzo 2014) or Ty1 and delta (δ) elements in *S. cerevisiae* (Da Silva and Srikrishnan 2012; Genbauffe et al. 1984), facilitate DNA movement into the genome.

2.5 Promoters

Optimal gene expression levels are critical to the success of both heterologous and endogenous pathway engineering efforts. In this vain, transcription of the gene to RNA (controlled by promoters) is the first step of this process. Therefore, understanding and controlling the regulation of transcription allow one to edit the rate of expression. The most common approach to change the rate of expression is by changing the promoter sequence driving the gene(s) of interest. Promoters are responsible for recruiting the necessary transcription machinery and initiating transcription elongation. Many different motifs are contained within a promoter sequence, and the complexity of these elements tends to scale with the complexity of the organism (more complex in higher eukaryotes). Together, these various elements aid the RNA polymerase to find promoter regions and open up the DNA to prepare it for transcription (Feklistov 2013). Promoters are typically categorized as either constitutive or inducible, depending on their activity and may be derived from endogenous, heterologous, or synthetic sources. Constitutive promoters are generally considered “on” under most conditions, while inducible promoters require a stimulus to change the mode of expression.

2.5.1 Constitutive Promoters

A strong emphasis has been placed in the field on the identification of strong, constitutive promoters (typically one of the first to be used to prototype a pathway of interest). For the case of bacteria, while several endogenous, constitutive promoters can be selected (usually ribosomal in nature), the vast majority of high strength promoters are based on phage sequences. There are several constitutive and native promoters in *E. coli* which have been studied for their expression levels under a variety of conditions (Singh 2014). As an example, Liang et al. characterized seven of these promoters: the spc ribosomal protein operon promoter P_{spc}, the β-lactamase gene promoter P_{bla}, the P_L promoter of phage λ, the replication control promoters P_{RNAI} and P_{RNAII}, and the P1 and P2 promoters of the *rrnB* ribosomal RNA operon (Liang et al. 1999). However, despite strong expression from each of these elements, there is still some interdependence on growth conditions.

Therefore, constitutive promoters derived from phages have been introduced into *E. coli* with great success. The T7 promoter, for example, is extremely successful in *E. coli* expression systems as it can lead to the production of target protein in excess of 40–50 % of the total cell protein (Baneyx 1999). To accomplish this, a T7 RNA polymerase must be introduced and expressed separately (some bacterial strains already have this in the genome). However, T7 RNA polymerase is unique in that it only recognizes T7 promoters, allowing for very specialized transcription and orthogonality between the native cell's transcription machinery. As such, this has become a very common expression system in bioprocesses (Rosano and Ceccarelli 2014; Berlec and Strukelj 2013; Baneyx 1999). Moreover, with advances in directed evolution, further orthogonal and distinct T7 promoter-polymerase pairs can further expand the scope of this system (Ellefson et al. 2014).

For the case of yeast, it is not possible to take motivation from phage, and thus most strong, constitutive promoters are those isolated from the yeast's genome (Da Silva and Srikrishnan 2012; Redden et al. 2014). Many of the native promoters characterized and used in yeast systems originate from the glycolytic pathway and have been shown to range in expression. As an example, Sun et al. characterized fourteen constitutive promoters, including the *ADH1*, *TEF1*, *TEF2*, and *GPD* promoters in the context of the most popular yeast expression vectors mentioned previously (Sun et al. 2012; Mumberg et al. 1995). Likewise, Partow et al. characterized seven additional yeast promoters, including the *TEF1, ADH1, TPI1, HXT7, TDH3, PGK1,* and *PYK1* promoters (Partow et al. 2010). In these characterizations, the relative strengths of the promoters were determined by measuring the amount of protein production enabled by each promoter sequence (Da Silva and Srikrishnan 2012; Redden et al. 2014; Sun et al. 2012; Partow et al. 2010). Unlike the bacterial system, though, these promoters are bounded to the levels of native transcription in yeast as they are derived from native promoters. Several reviews have been published that summarized the various strengths of common promoters for metabolic engineering applications (Da Silva and Srikrishnan 2012; Redden et al. 2014). However, synthetic approaches to promoter engineering are working to expand the toolbox of available promoters beyond those found natively in the cell.

2.5.2 Synthetic Promoters

Endogenous promoter sequences can be sufficient for certain applications of pathway engineering. However, their use in larger constructs with multiple genes is challenging because expression can be limited, there is a limited set of strong promoters, these elements may be subject to latent endogenous regulation, and their homology to the genome can prove to be unstable, especially in cells that perform homologous recombination (Dehli et al. 2012). Since the balancing of gene expression is important for pathway engineering (Ajikumar et al. 2010; Li et al. 2013), an increasing set of synthetic promoters and promoter libraries has begun to emerge. We briefly highlight a few of these approaches here.

Synthetic promoters can be designed in several ways. In one approach, error-prone polymerase chain reaction (PCR) mutagenesis is performed on a lead promoter sequence to introduce mutations at the promoter sequence level, thus leading to expression variation. This approach is generalizable across hosts and allows for new promoters to be identified with either increases or decreases in relative expression level (Redden et al. 2014; Alper et al. 2005; Nevoigt et al. 2006; Rajkumar and Maerkl 2012; Blazeck and Alper 2013). Typically, most members of the promoter library will be lower expression than that of the starting, lead promoter. However, libraries with a wide range of expression are of high utility when balancing enzymatic levels in a pathway as it is difficult to a priori determine the optimal expression level. Thus, these synthetic promoter libraries can be coupled with screening to identify the combinations leading to the best phenotype (Alper et al. 2005). Additionally, other synthetic promoter libraries have been used to find promoters that have better regulation than their wild-type counterparts (Nevoigt et al. 2006; Nevoigt et al. 2007). Such efforts have allowed minimal, synthetic promoters to be designed which improve upon native promoters or induce under culture conditions (Redden and Alper 2015).

The other approach to synthetic promoters is through rational design. In these instances, there is even less homology between the resulting synthetic promoter and endogenous promoters. For the case of yeast, within a promoter sequence, it is possible to dissect two major components: a core promoter and an upstream activating sequence (UAS). The core promoter typically contains the TATA box or other necessary transcriptional elements. The UAS, on the other hand, contains binding sites for transcription factors that aid in the recruitment of RNA polymerase to the DNA, but cannot initiate transcription by itself. In many cases, these two elements can be identified separately. Thus, a "hybrid promoter" approach is one in which the UAS sequence from one promoter is stitched to the core promoter of another. Using such an approach, high expression promoters have been made by combining highly active UAS elements to core promoters (Blazeck et al. 2011, 2012). Other promoter characteristics, like nucleosome occupancy, have also been engineered to make high-strength synthetic promoters (Curran et al. 2014). These results have led to pure de novo design of synthetic promoters that contain very little, if any, homology to endogenous elements. However, there is still much to be learned about the components that influence promoter strength, so the field of rationally designed promoters has room to expand the available toolbox for metabolic engineers.

2.5.3 Constitutive Versus Inducible Systems

High-strength, constitutive promoters which are always in the "on" state are extremely useful, especially for the first design of heterologous pathways. However, in order for an expression cassette to be successful at the industrial scale, the promoter must be both highly expressed and tightly regulated (Makrides 1996). In some cases, this desired regulation is constitutive and the promoters described above can suffice. However, the overexpression and accumulation of recombinant and heterologous protein in a cell can often be detrimental to the cell's productivity (Chou 2007). As such, it is often desirable to introduce tight regulation such that protein production can be turned on and off as desired. In order to do this, inducible promoters that are highly expressed have a low level of basal transcription, are transferable across strains, and have cheap and simple induction need to be used (Makrides 1996; Berlec and Strukelj 2013).

In *E. coli*, a wide variety of inducible promoters is available that can be induced via temperature (Chao et al. 2004; Wang et al. 2012), pH (Makrides 1996), or carbon source (Guzman et al. 1995). The most common and widely studied of these are those promoters derived from the bacterial *lac* operon (Graumann and Premstaller 2006). The *lac* operon and its promoter are known for their ability to respond to lactose or lactose analogs such as isopropyl-β-D-thio-galactoside (IPTG) and induce expression. However, the native *lac* promoter is relatively weak and becomes repressed in the presence of glucose, so derivatives, such as *lac*UV5, have been designed that are less sensitive to catabolites (Silverstone et al. 1970). Elements of this promoter have then been used in combination with the strong T7 RNA polymerase promoter, such as in the pET expression system, to achieve strong, inducible expression of a gene (Rosano and Ceccarelli 2014; Berlec and Strukelj 2013; Makoff and Oxer 1991). However, other expression systems, such the pBAD system based on the arabinose inducible promoter from the *araBAD* operon, are gaining popularity because the transcription rate scales with arabinose concentration (Guzman et al. 1995). This promoter is also repressed by glucose, though. Lastly, the strong λ promoter P_L has also been successfully used in a similar way by making a tryptophan inducible system (Mieschendahl et al. 1986).

Yeast systems also have a variety of inducible promoters used in expression applications, the *GAL* system of promoters being the most common (Johnston 1987). *GAL1*, *GAL7*, and *GAL10* are three promoters which exhibit induction in the presence of galactose instead of glucose (Bassel and Mortimer 1971). As such, these promoters have become very commonly used since galactose induction is relatively cheap and easy for cell culturing—however, such an element is not desirable when biomass sugars are used (i.e., when glucose will be present).

However, other yeast inducible promoters can also be used. The native gene *CUP1* from *Saccharomyces cerevisiae* is inducible by copper ions (Labbe and Thiele 1999). The promoter for *ADH2* is repressed in the presence of glucose (Price et al. 1990) and the *MET3* and *MET25* promoters are regulated by the presence of methionine (Sangsoda et al. 1984; Cherest et al. 1987). Yet, despite the availability, the *GAL* promoters remain the most commonly used because of their foldexpression change, their response time, and their ease of use in laboratory settings (Adams 1972). Thus, cheap, inducible, non-glucose repressed promoter elements need to be developed for fungal systems.

2.6 Translational Level Regulation

Transcriptional control by promoters can determine the level of overexpression or down regulation at the transcriptional level. After this point, additional bottlenecks/controls exist at the translational level. There are two primary points of regulation at the translational level: translation initiation and translation elongation. Translation initiation is regulated by the ability for tRNA, initiation factors, and rRNA to recognize and bind the messenger RNA and begin the translation process (Jackson et al. 2010). Translational elongation is regulated by the flux of the ribosome unit through the open reading frame (ORF) of the gene and the speed at which it can polymerize the polypeptide chain (Gorgoni et al. 2014). Steps can be taken to optimize both of these steps and increase the net expression of a gene. In addition, it is possible to manipulate mRNA half-life through a variety of techniques.

2.6.1 Translation Initiation

The first step in regulating translation occurs at the translation initiation level. Initiation occurs by first binding the ribosomal subunits and recruiting the necessary machinery to begin translation. In bacteria, most ribosomal binding is determined by the Shine–Dalgarno sequence on the 5′ untranslated region (UTR) of the mRNA transcript (Shine and Dalgarno 1975; Malys 2012). This sequence, spaced between 4-18 nucleotides upstream of the start codon, is responsible for recruiting and binding the ribosome near to the start codon. The efficiency of this sequence to recruit ribosomes has been shown to have an almost 1000-fold effect in protein expression (Malys 2012; Curry and Tomich 1988). As such, tools that predict these ribosomal binding sites (RBS) can be used to increase expression values. The RBS Calculator, for example, offers a way to predict and optimize RBS sequences in bacterial systems (Salis 2011). However, RBS strength is still influenced by the neighboring DNA sequence, so a strong Shine–Dalgarno sequence is not guaranteed to maximize translation initiation.

Ribosome binding in yeast, on the other hand, occurs further upstream of the start codon and recognizes the 5′ cap structure on mRNA transcripts. The ribosome then scans the transcript until it recognizes the start codon and Kozak sequence where it begins translation (Muller and Trachsel 1990; Yoon and Donahue 1992). In yeast, the optimum Kozak sequence had been identified to be 5′-AAUAAUGG-3′ for translation initiation (Hamilton et al. 1987). There have been studies, such as in the production of metallothionein III, where using an optimal consensus sequence in yeast has been shown to increase production (Wang et al. 1998). However, mutational analysis of this sequence showed that alternative sequences were sufficient for expression (Cigan et al. 1988), and because of the proposed scanning mechanism of the ribosome, the start codon closest to the 5′ end of the transcript is still usually favored (Kozak 2002).

2.6.2 Translation Elongation

Once translation initiates, the rate of expression will become limited by the rate at which ribosomes are able to move through the open reading frame (ORF) of the gene. Although this rate can be attributed to a couple factors, the most commonly engineered and optimized involves the availability of tRNA. Within the cell, the respective levels of each tRNA vary for each codon. This bias and availability of the correct tRNA for each codon can therefore become a limiting reagent in elongation (Rosano and Ceccarelli 2014; Gorgoni et al. 2014; Kane 1995). So far, there have been two common methods to combat this. One has been to codon optimize the genes to match the host organism. Due to the degeneracy of the genetic code (i.e., multiple tRNA coding for the same amino acid), targeted mutations in the DNA sequence can be used to swap rarer tRNA codons to those coding for more abundant tRNA without a change in amino acid sequence (Sorensen and Mortensen 2005; Gorgoni et al. 2014). Many algorithms exist to predict optimal codon optimization; however, these methods do not guarantee increased expression, and may lead to mRNA instability. An alternative approach has been to insert plasmids that increase the relative expression levels of the rare tRNA and increase their availability (Gorgoni et al. 2014; Kane 1995). Codon-optimized genes can be ordered for a variety of organisms online whereas strains overexpressing rare tRNA are available when codon optimization is not the preferred method of choice or fails to increase expression.

However, despite proper translation optimization, protein expression is still not guaranteed to increase. For example, the heterologous expression of P450 and P450 reductase in yeast saw an increase in protein production upon translation optimization (Batard et al. 2000; Gustafsson et al. 2004). Other systems, such as those used in the production of amorphadiene, saw little to no effect on protein output from codon bias optimization (Westfall et al. 2012). Therefore, the effect of translation optimization tends to be very transcript specific and reliant on multiple factors, such as relative transcript levels and mRNA structures (Gustafsson et al. 2004; Welch et al. 2009; Shah et al. 2013).

2.6.3 Terminator Design

The final method to increase the net rate of translation is to have more mRNA transcript available for the ribosomes. In the prior section, we described how transcript levels could be increased by using higher strength promoters and higher gene copy numbers. While these methods work to increase the production of transcript, the stability of these transcripts also determines their overall concentration in the cell. The selection of terminator can be used to influence net mRNA stability. Specifically, it had been found that found that mRNA stability changes depending on the terminator sequence and termination pathway (Abe and Aiba 1996). It was also seen that the termination efficiency varied greatly from one terminator sequence to another (Cambray et al. 2013). Therefore, one of the last ways to boost protein production is done by swapping the terminator sequence to a higher strength terminator (Yamanishi et al. 2013; Curran et al. 2013b). High strength terminator sequences have been shown to increase the net protein production by increasing the stability of the mRNA transcript (Yamanishi et al. 2013; Curran et al. 2013b; Yamanishi et al. 2011; Ito et al. 2013) and thus lead to more efficient pathways. Therefore, terminator sequences complement high-expression cassettes by leading to higher mRNA availability and bigger bursts of protein production with lower overall net transcriptional load. The development of terminator libraries and synthetic sequences is still very nascent and there is still much research to be done on the termination mechanism in cells. Thus, this is a growing area of synthetic biology research.

2.7 Synthetic Biology Tools

Many of the common bottlenecks to protein expression and pathway optimization have been traditionally addressed by stitching together a series of biological parts as described above. Specifically, elements including the plasmid, promoter sequence, ribosomal binding site, and terminator sequence are all parts that, when placed together, control the expression of a gene. However, as the library of these parts expands, interesting synthetic tools and paradigms are being developed that aid in the development of more sophisticated expression cassettes and control gene expression. Although this field is still relatively young, we will look at some of the synthetic biology tools currently available for optimizing gene expression pathways.

2.7.1 Logic Circuit Design

Gene cassettes and their parts are expected to have robust expression and tight regulation. For example, many natural biological processes rely on the ability to accurately time gene expression in response to the environmental conditions. There are some gene control elements capable of doing this, such as the inducible promoters described above, which allow expression to be modulated in response to a single environmental trigger. However, as pathways become more complicated, there is a need to make expression cassettes that have a variety of responses to multiple triggers. To do this, engineers have started developing and programming cellular gene circuits into cells in response to complex environmental patterns (Brophy and Voigt 2014).

The logic signaling circuits used in cells are often thought to be similar to the Boolean logic gates used in programming (Morris et al. 2010). For example, the enhancer regions used in gene circuits contain the binding sites for transcription factors that either silence or activate the expression of a gene (Amit 2012). Combinations of these binding sequence motifs make it possible to make Boolean logic AND gates and OR gates which activate transcription in response to which signals are present (Brophy and Voigt 2014; Morris et al. 2010; Ramalingam et al. 2009). In bacteria, the LacI, TetR, and CI repressors were some of the first to be used in the development of logic gates (Brophy and Voigt 2014; Ramalingam et al. 2009). However, the range of inputs has been expanded to include signals from pH, sugar content, metabolites, ligands, light, chemical pulses, or signaling proteins (Brophy and Voigt 2014; Morris et al. 2010; Wang and Buck 2012). Therefore, as this field expands, it will become possible to automate the production of these genetic circuits (Nielsen et al. 2016) and control each gene in a pathway based on the unique signals present (Brophy and Voigt 2014). Such control is expected to help optimize pathway expressions and increase regulation. Moreover, these approaches can lead to more versatile cells that can perform multiple functions all controlled by stimuli.

2.7.2 Synthetic Operons

Another approach to controlling the regulation of a series of genes has been through synthetic operons (sometimes referred to as "refactoring"). Bacterial cells have been studied for a number of years because of their ability to co-transcribe a number of genes as part of the same operon (Okuda et al. 2007). In an operon, multiple genes are transcribed onto the same mRNA and then translated in order to obtain multiple protein products from the same mRNA transcript (and thus controlled by a singular promoter element). The *lac* operon, for example, is one of the most well studied operons in *E. coli* and all bacterial systems (Lewis 2005, 2011). In it, multiple genes are transcribed in response to a single repressor protein that binds upstream of them on the DNA. Therefore, the use of these operons has started to grow in popularity for pathway engineering approaches.

Controlling the expression of multiple genes in a pathway can be challenging because the native regulation in the cell can be different or complicated for each gene. Synthetic operons allow one to circumvent this regulation by introducing these genes in operons that use well-characterized and regulated parts (Temme et al. 2012). For example, synthetic operons make it possible to refactor entire gene clusters onto a single operon that can be induced by different signals (Temme et al. 2012). The organization of these genes on the operon, the strength of the ribosome binding site, and posttranslational regulation can then be used to modulate the relative expression levels of each gene individually (Lim et al. 2011; Levin-Karp et al. 2013; Agnew and Pfleger 2011). This way, reactions can be coupled together and controlled with regulation that is well known and characterized for entire pathways (Temme et al. 2012; Lu and Ellington 2014; Matsumoto et al. 2011). This approach can be used to activate secondary metabolite production and remove regulation found in native pathways. In this regard, these approaches enable complete, synthetic control of a pathway.

2.7.3 Synthetic Feedback Loops

The balance of expression is important for the engineering of metabolic pathways, particularly when toxic or unproductive intermediates exist. For example, microbes can be used for biofuel production, but the fuels themselves are often toxic to cell growth (Dunlop et al. 2010). In traditional approaches, gene expression would have to be balanced through promoter engineering in order to prevent an overproduction of these toxic compounds. However, these approaches are laborious and cannot always account for fluctuations in environmental conditions.

Therefore, synthetic feedback loops have been used to add robustness to pathways and increase expression control. In a feedback loop, synthetic parts are designed that autoregulate their own expression in order to prevent toxic buildup in a cell. This results in a dynamic control over pathways rather than the static control afforded by constitutive promoters. When the expression of a protein increases its own expression, it is called a positive feedback loop; when a protein down regulates its own expression, it is a negative feedback loop. In the biofuel example, feedback from biofuel drove the expression of an efflux pump to keep the level of biofuel inside the cell from being prohibitive to growth (Dunlop et al. 2010; Harrison and Dunlop 2012). However, other times feedback loops are used to control transcriptional noise. Negative feedback loops allow gene expression to stay relatively static in response to environmental conditions that would otherwise skew their expression (Holtz and Keasling 2010). Other times, positive feedback loops allow biosensors to be more sensitive to external conditions by overexpressing when sensing metabolites (Kobori et al. 2013). Lastly, a combination of positive and negative feedback loops is used in synthetic circuit parts, such as oscillators, which allow cellular dynamics to be studied (Singh 2014; Brophy and Voigt 2014; Fung et al. 2005; Chen and Arkin 2012; Stricker et al. 2008; Elowitz and Leibler 2000; Atkinson et al. 2003; Yokobayashi et al. 2002; Gardner et al. 2000). However, the overarching idea of synthetic feedback loops is the ability to control a robust expression system that can adapt to environmental changes (either outside or inside the cell). These approaches can enable a flexible pathway that can respond to perturbation.

2.7.4 Metabolic Engineering Using Synthetic Biology Tools

Many of the synthetic biology tools listed above have applications in pathway and metabolic engineering by themselves. However, the combination of multiple approaches from the above has expedited and advanced metabolic engineering in a variety of applications. The production of taxadiene, an intermediate for the anti-cancer drug Taxol, is one example (Ajikumar et al. 2010). In this study, expression modules of a native pathway and a heterologous pathway were created using synthetic operons with a variety of promoter strengths. By collecting data under a variety of conditions, including plasmid-based and genomic-based expression, promoter strengths, and operon organization, they used multivariate analysis to obtain a 15,000-fold improvement in the production of taxadiene (Ajikumar et al. 2010). Whereas many studies would test these conditions separately, the use of multivariate analysis greatly improved the influences each variable also had on each other.

Another example is one in which the yield of biofuel production was increased threefold (Zhang et al. 2012). In this case, the balance of an intermediate metabolite in the biofuel production pathway was a key to maximizing the overall yield. Thus, these researchers used a biosensor that could change the protein production of the pathway in response to the intermediate's level. The pathway itself was split into multiple, synthetic modules that were then optimized underneath different promoter controls until an optimal expression level was obtained. Feedback loops were then used to increase and decrease gene expression levels and push metabolic flux into the production of fatty acids, increasing production to 28 % theoretical yield (Zhang et al. 2012).

These two examples emphasize that importance of finding the proper balance of enzymes needed for production. In both of these cases, varying the expression of gene modules, in combination with feedback loops controlling the level of intermediates, was able to achieve optimal production. As such, these synthetic tools offer very powerful advantages to a metabolic engineer, but there is still much to be discovered.

2.8 Conclusions

Many of the challenges in metabolic engineering come from the ability to properly balance the necessary proteins and enzymes in a pathway. It is often difficult to ab initio predict the optimal expression level for multiple genes in a pathway, so engineering has been largely empirical. However, throughout this chapter, we presented many of the tools available to a metabolic engineer in order to achieve fine control over the level of gene expression. Together, these various parts were designed to control the transcriptional level either at the transcription initiation phase or in the posttranscription and translation level.

The advances of synthetic biology, though, are certainly helping to create sophisticated control systems that can autoregulate pathways. Such systems and techniques are providing the promise of fully dynamic control in the cells, thus establishing a step toward continuous process optimization. The rapid screening of synthetic and pathway libraries is increasing the speed at which pathways can be optimized for overall productivity. Therefore, as we move forward and these tools are developed, we will further have our control over metabolism and develop better ways to rewire cells into cellular factories.

References

Abe H, Aiba H (1996) Differential contributions of two elements of rho-independent terminator to transcription termination and mRNA stabilization. Biochimie 78(11–12):1035–1042

Adams BG (1972) Induction of galactokinase in *Saccharomyces cerevisiae*: kinetics of induction and glucose effects. J Bacteriol 111(2):308–315

Agnew DE, Pfleger BF (2011) Optimization of synthetic operons using libraries of post-transcriptional regulatory elements. Methods Mol Biol 765:99–111

Ajikumar PK et al (2010) Isoprenoid pathway optimization for Taxol precursor overproduction in *Escherichia coli*. Science 330(6000):70–74

Alper H et al (2005) Tuning genetic control through promoter engineering. Proc Natl Acad Sci U S A 102(36):12678–12683

Amit R (2012) Towards synthetic gene circuits with enhancers: biology's multi-input integrators. Subcell Biochem 64:3–20

Anthony JR et al (2009) Optimization of the mevalonate-based isoprenoid biosynthetic pathway in *Escherichia coli* for production of the anti-malarial drug precursor amorpha-4,11-diene. Metab Eng 11(1):13–19

Atkinson MR et al (2003) Development of genetic circuitry exhibiting toggle switch or oscillatory behavior in *Escherichia coli*. Cell 113(5):597–607

Baneyx F (1999) Recombinant protein expression in *Escherichia coli*. Curr Opin Biotechnol 10 (5):411–421

Bassel J, Mortimer R (1971) Genetic order of the galactose structural genes in *Saccharomyces cerevisiae*. J Bacteriol 108(1):179–183

Batard Y et al (2000) Increasing expression of P450 and P450-reductase proteins from monocots in heterologous systems. Arch Biochem Biophys 379(1):161–169

Berlec A, Strukelj B (2013) Current state and recent advances in biopharmaceutical production in *Escherichia coli*, yeasts and mammalian cells. J Ind Microbiol Biotechnol 40(3–4):257–274

Blazeck J, Alper HS (2013) Promoter engineering: recent advances in controlling transcription at the most fundamental level. Biotechnol J 8(1):46–58

Blazeck J et al (2011) Tuning gene expression in *Yarrowia lipolytica* by a hybrid promoter approach. Appl Environ Microbiol 77(22):7905–7914

Blazeck J et al (2012) Controlling promoter strength and regulation in *Saccharomyces cerevisiae* using synthetic hybrid promoters. Biotechnol Bioeng 109(11):2884–2895

Blazeck J et al (2014a) Metabolic engineering of *Saccharomyces cerevisiae* for itaconic acid production. Appl Microbiol Biotechnol 98(19):8155–8164

Blazeck J et al (2014b) Harnessing Yarrowia lipolytica lipogenesis to create a platform for lipid and biofuel production. Nat Commun 5:3131

Bolivar F et al (1976) Construction and characterization of new cloning vehicles. II. A multipurpose cloning system. Gene 2(2):95–113

Brophy JA, Voigt CA (2014) Principles of genetic circuit design. Nat Methods 11(5):508–520

Cambray G et al (2013) Measurement and modeling of intrinsic transcription terminators. Nucleic Acids Res 41(9):5139–5148

Camps M (2009) Modulation of ColE1-like plasmid replication for recombinant gene expression. Recent Pat DNA Gene Sequences 4(1):58–73

Celik E, Calik P (2012) Production of recombinant proteins by yeast cells. Biotechnol Adv 30 (5):1108–1118

Chang AC, Cohen SN (1978) Construction and characterization of amplifiable multicopy DNA cloning vehicles derived from the P15A cryptic miniplasmid. J Bacteriol 134(3):1141–1156

Chao YP, Wen CS, Wang JY (2004) A facile and efficient method to achieve LacZ overproduction by the expression vector carrying the thermoregulated promoter and plasmid copy number. Biotechnol Prog 20(2):420–425

Chen R (2012) Bacterial expression systems for recombinant protein production: *E. coli* and beyond. Biotechnol Adv 30(5):1102–1107

Chen D, Arkin AP (2012) Sequestration-based bistability enables tuning of the switching boundaries and design of a latch. Mol Syst Biol 8:620

Chen X et al (2013) Metabolic engineering of Escherichia coli: a sustainable industrial platform for bio-based chemical production. Biotechnol Adv 31(8):1200–1223

Cherest H, Kerjan P, Surdin-Kerjan Y (1987) The *Saccharomyces cerevisiae* MET3 gene: nucleotide sequence and relationship of the 5' non-coding region to that of MET25. Molecular & general genetics: MGG 210(2):307–313

Chou CP (2007) Engineering cell physiology to enhance recombinant protein production in *Escherichia coli*. Appl Microbiol Biotechnol 76(3):521–532

Cigan AM, Pabich EK, Donahue TF (1988) Mutational analysis of the HIS4 translational initiator region in *Saccharomyces cerevisiae*. Mol Cell Biol 8(7):2964–2975

Clarke L, Carbon J (1980) Isolation of a yeast centromere and construction of functional small circular chromosomes. Nature 287(5782):504–509

Crook NC, Schmitz AC, Alper HS (2014) Optimization of a yeast RNA interference system for controlling gene expression and enabling rapid metabolic engineering. ACS Synth Biol 3 (5):307–313

Curran KA et al (2013a) Metabolic engineering of muconic acid production in *Saccharomyces cerevisiae*. Metab Eng 15:55–66

Curran KA et al (2013b) Use of expression-enhancing terminators in *Saccharomyces cerevisiae* to increase mRNA half-life and improve gene expression control for metabolic engineering applications. Metab Eng 19:88–97

Curran KA et al (2014) Design of synthetic yeast promoters via tuning of nucleosome architecture. Nat Commun 5:4002

Curry KA, Tomich CS (1988) Effect of ribosome binding site on gene expression in *Escherichia coli*. DNA 7(3):173–179

Da Silva NA, Srikrishnan S (2012) Introduction and expression of genes for metabolic engineering applications in *Saccharomyces cerevisiae*. FEMS Yeast Res 12(2):197–214

Dalton AC, Barton WA (2014) Over-expression of secreted proteins from mammalian cell lines. Protein Sci 23(5):517–525
Datsenko KA, Wanner BL (2000) One-step inactivation of chromosomal genes in *Escherichia coli* K-12 using PCR products. Proc Natl Acad Sci USA 97(12):6640–6645
Dehli T, Solem C, Jensen PR (2012) Tunable promoters in synthetic and systems biology. Subcell Biochem 64:181–201
del Solar G, Espinosa M (2000) Plasmid copy number control: an ever-growing story. Mol Microbiol 37(3):492–500
del Solar G et al (1998) Replication and control of circular bacterial plasmids. Microbiol Mol Biol Rev 62(2):434–464
DiCarlo JE et al (2013) Genome engineering in *Saccharomyces cerevisiae* using CRISPR-Cas systems. Nucleic Acids Res 41(7):4336–4343
Dunlop MJ, Keasling JD, Mukhopadhyay A (2010) A model for improving microbial biofuel production using a synthetic feedback loop. Syst Synth Biol 4(2):95–104
Ellefson JW et al (2014) Directed evolution of genetic parts and circuits by compartmentalized partnered replication. Nat Biotechnol 32(1):97–101
Elowitz MB, Leibler S (2000) A synthetic oscillatory network of transcriptional regulators. Nature 403(6767):335–338
Feklistov A (2013) RNA polymerase: in search of promoters. Ann N Y Acad Sci 1293:25–32
Fung E et al (2005) A synthetic gene-metabolic oscillator. Nature 435(7038):118–122
Futcher B, Carbon J (1986) Toxic effects of excess cloned centromeres. Mol Cell Biol 6(6):2213–2222
Gardner TS, Cantor CR, Collins JJ (2000) Construction of a genetic toggle switch in *Escherichia coli*. Nature 403(6767):339–342
Genbauffe FS, Chisholm GE, Cooper TG (1984) Tau, sigma, and delta. A family of repeated elements in yeast. J Biol Chem 259(16):10518–10525
Giaever G, Nislow C (2014) The yeast deletion collection: a decade of functional genomics. Genetics 197(2):451–465
Giaever G et al (2002) Functional profiling of the *Saccharomyces cerevisiae* genome. Nature 418 (6896):387–391
Goh S, Good L (2008) Plasmid selection in *Escherichia coli* using an endogenous essential gene marker. BMC Biotechnol 8:61
Gorgoni B et al (2014) Controlling translation elongation efficiency: tRNA regulation of ribosome flux on the mRNA. Biochem Soc Trans 42(1):160–165
Graumann K, Premstaller A (2006) Manufacturing of recombinant therapeutic proteins in microbial systems. Biotechnol J 1(2):164–186
Gustafsson C, Govindarajan S, Minshull J (2004) Codon bias and heterologous protein expression. Trends Biotechnol 22(7):346–353
Guzman LM et al (1995) Tight regulation, modulation, and high-level expression by vectors containing the arabinose PBAD promoter. J Bacteriol 177(14):4121–4130
Hägg P et al (2004) A host/plasmid system that is not dependent on antibiotics and antibiotic resistance genes for stable plasmid maintenance in *Escherichia coli*. J Biotechnol 111(1):17–30
Hamilton R, Watanabe CK, de Boer HA (1987) Compilation and comparison of the sequence context around the AUG startcodons in *Saccharomyces cerevisiae* mRNAs. Nucleic Acids Res 15(8):3581–3593
Harrison ME, Dunlop MJ (2012) Synthetic feedback loop model for increasing microbial biofuel production using a biosensor. Front Microbiol 3:360
Holladay JE et al (2007) Top value-added chemicals from biomass: results of screening for potential candidates from biorefinery lignin. Pacific Northwest National Laboratory
Holtz WJ, Keasling JD (2010) Engineering static and dynamic control of synthetic pathways. Cell 140(1):19–23
Ito Y et al (2013) Characterization of five terminator regions that increase the protein yield of a transgene in *Saccharomyces cerevisiae*. J Biotechnol 168(4):486–492

Jackson RJ, Hellen CU, Pestova TV (2010) The mechanism of eukaryotic translation initiation and principles of its regulation. Nat Rev Mol Cell Biol 11(2):113–127

Jiang W et al (2013) RNA-guided editing of bacterial genomes using CRISPR-Cas systems. Nat Biotechnol 31(3):233–239

Johnston M (1987) A model fungal gene regulatory mechanism: the GAL genes of *Saccharomyces cerevisiae*. Microbiol Rev 51(4):458–476

Jones KL, Kim SW, Keasling JD (2000) Low-copy plasmids can perform as well as or better than high-copy plasmids for metabolic engineering of bacteria. Metab Eng 2(4):328–338

Kane JF (1995) Effects of rare codon clusters on high-level expression of heterologous proteins in *Escherichia coli*. Curr Opin Biotechnol 6(5):494–500

Karim AS, Curran KA, Alper HS (2013) Characterization of plasmid burden and copy number in *Saccharomyces cerevisiae* for optimization of metabolic engineering applications. FEMS Yeast Res 13(1):107–116

Kobori S et al (2013) A controllable gene expression system in liposomes that includes a positive feedback loop. Mol BioSyst 9(6):1282–1285

Kozak M (2002) Pushing the limits of the scanning mechanism for initiation of translation. Gene 299(1–2):1–34

Kroll J et al (2009) Establishment of a novel anabolism-based addiction system with an artificially introduced mevalonate pathway: Complete stabilization of plasmids as universal application in white biotechnology. Metab Eng 11(3):168–177

Kroll J et al (2010) Plasmid addiction systems: perspectives and applications in biotechnology. Microb Biotechnol 3(6):634–657

Kroll J, Klinter S, Steinbuchel A (2011) A novel plasmid addiction system for large-scale production of cyanophycin in *Escherichia coli* using mineral salts medium. Appl Microbiol Biotechnol 89(3):593–604

Labbe S, Thiele DJ (1999) Copper ion inducible and repressible promoter systems in yeast. Methods Enzymol 306:145–153

Lee C et al (2006) Absolute and relative QPCR quantification of plasmid copy number in *Escherichia coli*. J Biotechnol 123(3):273–280

Levin-Karp A et al (2013) Quantifying translational coupling in *E. coli* synthetic operons using RBS modulation and fluorescent reporters. ACS Synth. Biol 2(6):327–336

Lewis M (2005) The lac repressor. C R Biol 328(6):521–548

Lewis M (2011) A tale of two repressors. J Mol Biol 409(1):14–27

Li Z et al (2013) Direct and efficient xylitol production from xylan by *Saccharomyces cerevisiae* through transcriptional level and fermentation processing optimizations. Bioresour Technol 149:413–419

Liang S et al (1999) Activities of constitutive promoters in *Escherichia coli*. J Mol Biol 292(1): 19–37

Lim HN, Lee Y, Hussein R (2011) Fundamental relationship between operon organization and gene expression. Proc Natl Acad Sci USA 108(26):10626–10631

Liu L et al (2013a) How to achieve high-level expression of microbial enzymes: strategies and perspectives. Bioengineered 4(4):212–223

Liu L, Redden H, Alper HS (2013b) Frontiers of yeast metabolic engineering: diversifying beyond ethanol and Saccharomyces. Curr Opin Biotechnol 24(6):1023–1030

Lu WC, Ellington AD (2014) Design and selection of a synthetic operon. ACS Synth Biol 3 (6):410–415

Madyagol M et al (2011) Gene replacement techniques for *Escherichia coli* genome modification. Folia Microbiol (Praha) 56(3):253–263

Makoff AJ, Oxer MD (1991) High level heterologous expression in *E. coli* using mutant forms of the lac promoter. Nucleic Acids Res 19(9):2417–2421

Makrides SC (1996) Strategies for achieving high-level expression of genes in *Escherichia coli*. Microbiol Rev 60(3):512–538

Malys N (2012) Shine-Dalgarno sequence of bacteriophage T4: GAGG prevails in early genes. Mol Biol Rep 39(1):33–39

Martinez JL et al (2012) Pharmaceutical protein production by yeast: towards production of human blood proteins by microbial fermentation. Curr Opin Biotechnol 23(6):965–971

Matsumoto Y et al (2011) Bacterial cells carrying synthetic dual-function operon survived starvation. J Biomed Biotechnol 2011:489265

Mead DJ, Gardner DC, Oliver SG (1986) The yeast 2 micron plasmid: strategies for the survival of a selfish DNA. Mol Gen Genet 205(3):417–421

Mieschendahl M, Petri T, Hanggi U (1986) A Novel prophage independent Trp regulated lambda-Pl expression system. Bio-Technology 4(9):802–809

Minton NP (1984) Improved plasmid vectors for the isolation of translational lac gene fusions. Gene 31(1–3):269–273

Morris MK et al (2010) Logic-based models for the analysis of cell signaling networks. Biochemistry 49(15):3216–3224

Muller PP, Trachsel H (1990) Translation and regulation of translation in the yeast *Saccharomyces cerevisiae*. Eur J Biochem 191(2):257–261

Mumberg D, Muller R, Funk M (1995) Yeast vectors for the controlled expression of heterologous proteins in different genetic backgrounds. Gene 156(1):119–122

Nevoigt E et al (2006) Engineering of promoter replacement cassettes for fine-tuning of gene expression in *Saccharomyces cerevisiae*. Appl Environ Microbiol 72(8):5266–5273

Nevoigt E et al (2007) Engineering promoter regulation. Biotechnol Bioeng 96(3):550–558

Nielsen J (2013) Production of biopharmaceutical proteins by yeast: advances through metabolic engineering. Bioengineered 4(4):207–211

Nielsen AA et al (2016) Genetic circuit design automation. Science 352(6281):aac7341

Nordstrom K (2006) Plasmid R1–replication and its control. Plasmid 55(1):1–26

Okuda S et al (2007) Characterization of relationships between transcriptional units and operon structures in *Bacillus subtilis* and *Escherichia coli*. BMC Genom 8:48

Partow S et al (2010) Characterization of different promoters for designing a new expression vector in *Saccharomyces cerevisiae*. Yeast 27(11):955–964

Peubez I et al (2010) Antibiotic-free selection in *E. coli*: new considerations for optimal design and improved production. Microb Cell Fact 9:65

Price VL et al (1990) Expression of heterologous proteins in *Saccharomyces cerevisiae* using the ADH2 promoter. Methods Enzymol 185:308–318

Rajkumar AS, Maerkl SJ (2012) Rapid synthesis of defined eukaryotic promoter libraries. ACS Synth Biol 1(10):483–490

Ramalingam KI et al (2009) Forward engineering of synthetic bio-logical AND gates. Biochem Eng J 47(1–3):38–47

Redden H, Alper HS (2015) The development and characterization of synthetic minimal yeast promoters. Nat Commun 6:7810

Redden H, Morse N, Alper HS (2014) The synthetic biology toolbox for tuning gene expression in yeast. FEMS Yeast Res

Ro DK et al (2006) Production of the antimalarial drug precursor artemisinic acid in engineered yeast. Nature 440(7086):7940–7943

Rosano GL, Ceccarelli EA (2014) Recombinant protein expression in *Escherichia coli*: advances and challenges. Front Microbiol 5:172

Salis HM (2011) The ribosome binding site calculator. Methods Enzymol 498:19–42

Sanchez S, Demain A (2012) Special issue on the production of recombinant proteins. Biotechnol Adv 30(5):1100–1101

Sander JD, Joung JK (2014) CRISPR-Cas systems for editing, regulating and targeting genomes. Nat Biotechnol 32(4):347–355

Sangsoda S, Cherest H, Surdin-Kerjan Y (1984) The expression of the MET25 gene of *Saccharomyces cerevisiae* is regulated transcriptionally. Molecular & general genetics: MGG 200(3):407–414

Sauer B (1987) Functional expression of the cre-lox site-specific recombination system in the yeast *Saccharomyces cerevisiae*. Mol Cell Biol 7(6):2087–2096

Shah P et al (2013) Rate-limiting steps in yeast protein translation. Cell 153(7):1589–1601

Shine J, Dalgarno L (1975) Determinant of cistron specificity in bacterial ribosomes. Nature 254 (5495):34–38

Silva-Rocha R, de Lorenzo V (2014) Chromosomal integration of transcriptional fusions. Methods Mol Biol 1149:479–489

Silverstone AE, Arditti RR, Magasanik B (1970) Catabolite-insensitive revertants of lac promoter mutants. Proc Natl Acad Sci U S A 66(3):773–779

Singh V (2014) Recent advancements in synthetic biology: current status and challenges. Gene 535(1):1–11

Sorensen HP, Mortensen KK (2005) Advanced genetic strategies for recombinant protein expression in *Escherichia coli*. J Biotechnol 115(2):113–128

Stoker NG, Fairweather NF, Spratt BG (1982) Versatile low-copy-number plasmid vectors for cloning in *Escherichia coli*. Gene 18(3):335–341

Stricker J et al (2008) A fast, robust and tunable synthetic gene oscillator. Nature 456(7221):516–519

Suess B et al (2012) Aptamer-regulated expression of essential genes in yeast. Methods Mol Biol 824:381–391

Sun J et al (2012) Cloning and characterization of a panel of constitutive promoters for applications in pathway engineering in *Saccharomyces cerevisiae*. Biotechnol Bioeng 109 (8):2082–2092

Taxis C, Knop M (2006) System of centromeric, episomal, and integrative vectors based on drug resistance markers for *Saccharomyces cerevisiae*. Biotechniques 40(1):73–78

Temme K, Zhao D, Voigt CA (2012) Refactoring the nitrogen fixation gene cluster from *Klebsiella oxytoca*. Proc Natl Acad Sci U S A 109(18):7085–7090

Terpe K (2006) Overview of bacterial expression systems for heterologous protein production: from molecular and biochemical fundamentals to commercial systems. Appl Microbiol Biotechnol 72(2):211–222

Voss I, Steinbuchel A (2006) Application of a KDPG-aldolase gene-dependent addiction system for enhanced production of cyanophycin in *Ralstonia eutropha* strain H16. Metab Eng 8 (1):66–78

Walsh G (2010) Biopharmaceutical benchmarks 2010. Nat Biotechnol 28(9):917–924

Walsh G (2014) Biopharmaceutical benchmarks 2014. Nat Biotechnol 32(10):992–1000

Wang B, Buck M (2012) Customizing cell signaling using engineered genetic logic circuits. Trends Microbiol 20(8):376–384

Wang RF, Kushner SR (1991) Construction of versatile low-copy-number vectors for cloning, sequencing and gene expression in *Escherichia coli*. Gene 100:195–199

Wang SH, Shih YH, Lin LY (1998) Yeast consensus initiator sequence markedly enhances the synthesis of metallothionein III in *Saccharomyces cerevisiae*. Biotechnol Lett 20(1):9–13

Wang Z et al (2012) Expression and production of recombinant cis-epoxysuccinate hydrolase in *Escherichia coli* under the control of temperature-dependent promoter. J Biotechnol 162 (2–3):232–236

Welch M et al (2009) You're one in a googol: optimizing genes for protein expression. J R Soc Interface 6(Suppl 4):S467–S476

Werpy T et al (2004) Top value added chemicals from biomass: results of screening for potential candidates from sugars and synthesis gas. [U.S. Department of Energy [Office of] Energy Efficiency and Renewable Energy

Westfall PJ et al (2012) Production of amorphadiene in yeast, and its conversion to dihydroartemisinic acid, precursor to the antimalarial agent artemisinin. Proc Natl Acad Sci U S A 109(3):E111–E118

Winzeler EA et al (1999) Functional characterization of the *S. cerevisiae* genome by gene deletion and parallel analysis. Science 285(5429):5901–5906

Wu X et al (2014) Genome-wide binding of the CRISPR endonuclease Cas9 in mammalian cells. Nat Biotechnol 32(7):670–676

Yamanishi M, Katahira S, Matsuyama T (2011) TPS1 terminator increases mRNA and protein yield in a *Saccharomyces cerevisiae* expression system. Biosci Biotechnol Biochem 75 (11):2234–2236

Yamanishi M et al (2013) A genome-wide activity assessment of terminator regions in *Saccharomyces cerevisiae* provides a "terminatome" toolbox. ACS Synth Biol 2(6):337–347

Yao YF et al (2013) Metabolic engineering of *Escherichia coli* for production of salvianic acid A via an artificial biosynthetic pathway. Metab Eng 19:79–87

Yokobayashi Y, Weiss R, Arnold FH (2002) Directed evolution of a genetic circuit. Proc Natl Acad Sci USA 99(26):16587–16591

Yoon H, Donahue TF (1992) Control of translation initiation in *Saccharomyces cerevisiae*. Mol Microbiol 6(11):1413–1419

Zhang F, Carothers JM, Keasling JD (2012) Design of a dynamic sensor-regulator system for production of chemicals and fuels derived from fatty acids. Nat Biotechnol 30(4):354–359

Zhu J (2012) Mammalian cell protein expression for biopharmaceutical production. Biotechnol Adv 30(5):1158–1170

Zielenkiewicz U, Cegłowski P (2001) Mechanisms of plasmid stable maintenance with special focus on plasmid addiction systems. Acta Biochim Pol 48(4):1003–1023

Chapter 3
Heterologous Pathway Engineering

Wei Niu, Jiantao Guo and Steve Van Dien

Abstract Heterologous pathways encompass both natural and artificial biosynthetic routes. Expression of a heterologous pathway expands the molecular diversity that can be realized by the host organism. The engineering efforts often benefit from a well-established technological platform of the host. This chapter discusses recent progress and key challenges in implementing heterologous pathways. Major topics include enzyme discovery for artificial pathway assembly, methods for studying and tuning the performance of a pathway, and examples of heterologous pathway engineering.

Keywords Heterologous pathway · Biosynthesis · Enzyme · Metabolic engineering

3.1 Heterologous Biosynthetic Pathways

Traditionally, microbial syntheses are established and improved through strain breeding that entails chemical mutagenesis followed by phenotype screening in order to obtain a microorganism that can synthesize the desired compounds in high concentration and/or yield. Whereas this approach has been reasonably successful, the types of compounds synthesized are generally restricted to molecules that are part of a given microbe's natural metabolic network. Owing to scientific discoveries and technological progress in molecular biology, biochemistry and microbiology, and the emergence of new research areas including synthetic biology and systems biology,

W. Niu (✉)
Department of Chemical and Biomolecular Engineering, University of Nebraska–Lincoln, Lincoln, NE 68588, USA
e-mail: wniu2@unl.edu

J. Guo
Department of Chemistry, University of Nebraska–Lincoln, Lincoln, NE 68588, USA

S. Van Dien
Genomatica, Inc., San Diego, CA 92126, USA

S. Van Dien (ed.), *Metabolic Engineering for Bioprocess Commercialization*,
DOI 10.1007/978-3-319-41966-4_3

microbes can now be routinely manipulated to express a heterologous biosynthetic pathway in order to synthesize nonnative metabolites. Such heterologous pathways include naturally existing and artificially assembled biosynthetic pathways.

Expression of a natural biosynthetic pathway in a heterologous host strain may be preferred for a variety of reasons, such as poor growth properties of the natural host or the availability of its genetic engineering tools. The use of recombinant yeast strains to achieve large-scale economical production of artemisinic acid serves as an excellent example in this case (Ro et al. 2006; Paddon and Keasling 2014). Artemisinic acid is the biosynthetic precursor of antimalarial drug artemisinin, which is a sesquiterpene lactone accumulated in *Artemisia annua*. Despite its potency against chloroquine-resistant strain of *Plasmodium falciparum*, a wide access to artemisinin-containing drug to patients in disease-plagued developing countries is hindered by the cost of the plant-derived molecule. Due to challenges in manipulating secondary metabolite biosynthesis in plant, *Saccharomyces cerevisiae* was eventually chosen as the heterologous host for the production of artemisinic acid, which is further chemically converted to artemisinin (Fig. 3.1). After nearly ten years of research that focused on metabolic and process engineering, a yeast strain with the reconstituted plant biosynthetic pathway was able to produce artemisinic acid at 25 g/L (Paddon et al. 2013).

Devising a sequence of chemical reactions to achieve the total synthesis of a target molecule is a general practice in Organic Chemistry (Corey 1988). Similar rational was introduced into microbial syntheses by recruiting enzymatic reactions from multiple genetic origins in order to assemble an artificial biosynthetic pathway to produce the desired product (Draths and Frost 1994; Niu et al. 2003; Yim et al. 2011). Living systems possess enormous synthetic power to convert organic (such as sugar) and inorganic (such as CO_2) matters into metabolites with diverse structural and functional properties. One can envision a multidimensional biosynthetic network that

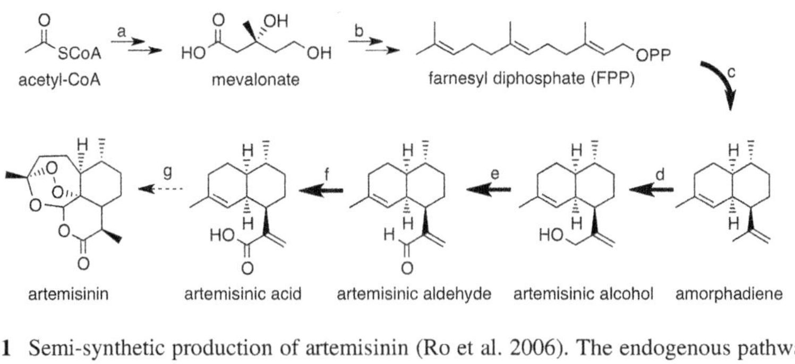

Fig. 3.1 Semi-synthetic production of artemisinin (Ro et al. 2006). The endogenous pathways of *S. cerevisiae* CEN.PK2 was engineered to increase the carbon flux into FPP biosynthesis. The *A. annua* artemisinic acid biosynthetic pathway (*arrows in bold face*) was reconstituted in engineered *S. cerevisiae*. Biosynthesized artemisinic acid was chemically converted into artemisinin. Enzymes and pathways (encoding genes): **a** mevalonate pathway; **b** FPP biosynthesis; **c** amorphadiene synthase (*ADS*); **d** cytochrome P450 (*CYP71AV1*), cytochrome P450 reductase (*CPR1*), and cytochrome b_5 (*CYB5*); **e** artemisinic alcohol dehydrogenase (*ADH1*); **f** artemisinic aldehyde dehydrogenase (*ALDH1*); **g** chemical synthesis

includes all known enzymatic reactions with each metabolite being represented as a dot and corresponding enzyme as a connecting line. Within this network, biosynthetic pathways can be rewired, extended, and created, and natural metabolites can be produced through artificially designed routes using methods described in Chap. 2. The promiscuous substrate specificity and, sometimes, catalytic activity of natural enzymes further enrich the complexity of the network by expanding the structural diversity of potential small-molecule products. In addition, a promiscuous enzyme activity can be optimized through enzyme engineering (Minshull and Stemmer 1999; Schmidt-Dannert 2001; Arnold 2001). New chemistry can be implemented by novel catalytic sites that are de novo constructed using computation-guided design (Röthlisberger et al. 2008; Siegel et al. 2010; Richter et al. 2011). Consequently, artificial pathways can be created to achieve the biosynthesis of nonnatural metabolites. Recent application of artificially designed biosynthetic pathways in microbial synthesis is propelled by its promise in solving societal challenges in energy, sustainability, and health care through utilizing renewable starting materials in the production of fuel molecules, commodity chemicals, and valuable pharmaceutical precursors. The focus of this chapter is to discuss recent progress and key challenges in implementing heterologous pathways (designed as described in Chap.2); in particular, a focus on artificial pathways. We also intend to provide information on resources that can facilitate the engineering process.

3.2 Enzyme Discovery

Although the heterologous expression of a naturally existing pathway can benefit from the known genetic information of pathway enzymes, quite often an optimal pathway performance requires enzymes with better physical or kinetic properties. In these cases, isozymes need to be examined and the best candidate needs to be integrated into the existing pathway. On the other hand, an artificial pathway often enlists enzyme activity on a nonnatural molecule. The identification of enzyme candidates that are functional in such a reaction is the initial challenging step toward building a feasible artificial biosynthetic pathway.

3.2.1 To Find the Most Promising Enzyme Candidates

Metagenomic library screening is proven to be an effective method for the discovery of novel catalytic activities (Handelsman et al. 1998). By screening hundreds of thousands of expression clones that are constructed from microbial natural habitat (e.g., soil or fresh water), one can identify desirable enzyme activity. Subsequent DNA sequencing, sub-cloning, and biochemical characterization are used to reveal the gene encoding that particular activity. Methods on the construction and the screening of microbial metagenomic library are extensively reviewed in literatures (Daniel 2002; Simon and Daniel 2011; DeLong 2013). Therefore, they will not be covered in this article. A second enzyme discovery approach requires intensive database search followed by wet lab biochemical characterization. Database search zooms in on a subset of known enzymes that most likely can catalyze the desirable reaction based on the reaction type and substrates similarity. Due to the large number of known enzymatic reactions, this initial step can be a taunting task when it is completely relied on manual search. One can use pathway design algorithms in conjunction with the database search as an imbedded function (Medema et al. 2012). Alternately, once a natural reaction that is similar to the desired one is identified, a recently developed EC-BLAST algorithm can simplify the search by querying ~6000 curated reactions and providing additional enzyme candidates based on bond change, reaction center, and reaction structure similarity (Rahman et al. 2014). The ENZYME repository of the Expert Protein Analysis System (ExPASy), which is operated by the Swiss Institute of Bioinformatics, is one excellent resource portal for the genetic and biochemical information of known enzymes. The database categorizes enzymes using the Enzyme Commission (EC) number. It cross-references with other major enzyme and genomic databases, including the Kyoto Encyclopedia of Genes and Genomes (KEGG), the BRaunschweig ENzyme DAtabase (BRENDA), the RCSB Protein Data Bank (PDB), and Metacyc database.

3.2.2 Functional Expression of Enzyme Candidates

With a number of candidate enzymes identified from the database search in hand, functional expression of the protein in the designated heterologous host microorganism is required for biochemical characterization. The two major factors that affect the functional expression are gene translation constraints and enzyme functionality constraints.

The codon usage bias across different species plays a major role in limiting the expression level of a heterologous gene. It is generally accepted that the speed of protein synthesis is largely determined by the cellular concentrations of tRNAs (Andersson and Kurland 1990). A heterologous gene that contains large number of rare codons (of which the cognate tRNAs have extremely low abundance) is often poorly expressed. Optimization of the codon usage of a heterologous gene generally increases the translational speed and results in higher level of protein expression (Gustafsson et al. 2004). Bioinformatics tools are available for the comparison of codon usages frequency between a heterologous gene and an expression host. Extensive genome sequencing projects also led to the establishment of databases that provide statistical information of codon usage in many organisms (Nakamura et al. 2000). In addition to single codon usage, the codon pair bias has also been taken into consideration in the gene optimization process (Gutman and Hatfield 1989). We recommend several excellent review articles on the subjects of codon bias, gene design, and protein expression (Welch et al. 2011; Plotkin and Kudla

2011; Novoa and de Pouplans 2012). Customized syntheses of codon-optimized genes are becoming routine services provided by many commercial companies. Besides codon optimization of the entire coding gene, other approaches can also be applied to boost the expression level of heterologous genes. Early research showed that the coding sequence of the N-terminus of a protein is critical to its expression level (Nassal et al. 1987). By partially optimizing the codon usage in this region or fusing a short N-terminal tag to the target gene, the protein expression can be efficiently increased. Alternatively, to satisfy the requirement of rare tRNAs for the translation of nonoptimized gene, the corresponding tRNA-encoding genes can be co-expressed. In the case when codon usage optimization does not result in satisfactory protein expression, other factors that affect protein expression including mRNA structure and mRNA stability of the gene should be examined. Additionally, intrinsic properties of the target protein are critical to its expression. In general, transmembrane proteins or proteins that are toxic to the host cannot be expressed well. The strategies discussed here and in this chapter can be applied to improve expression of heterologous proteins; however, the possibility remains that the gene will never express well in the host. Researchers must decide how much effort to devote on expressing a troublesome candidate as opposed to searching for and testing other candidates.

Quite often the function of an enzyme requires special cofactor, prosthetic group, or posttranslational modification. When these essential factors cannot be synthesized or installed by the heterologous host, desired proteins or pathways that serve the auxiliary functions need to be co-expressed. One example is the implementation of polyketide biosynthesis pathway in *E. coli*. Due to the narrow substrate specificity of *E. coli* native phosphopantetheinyl transferases, a *Bacillus subtilis* Sfp protein is often used to activate the apo polyketide synthase (Quadri et al. 1998). Another example is the heterologous expression of the *Clostridium acetobutylicum* butylryl-CoA dehydrogenase, which functions in the 1-butanol biosynthesis pathway. The enzyme requires the co-expression of a specific flavoprotein (EntAB) as the redox partner in the catalysis (Bounton et al. 1996). In addition, certain enzymes suffer mechanism-based inactivation and require a chaperone protein to regain the catalytic activities. The vitamin B_{12}-dependent glycerol dehydratase (GDH), which functions in the 1,3-propanediol biosynthetic pathway from glycerol, undergoes suicide radical reaction with substrate glycerol and leads to a bound incompetent cofactor (Nakamura and Whited 2003). The GDH activase is needed to repair the enzyme by replacing inactive cofactor with coenzyme B_{12} (Mori et al. 1997).

3.2.3 Characterization of Enzyme Candidates

Biochemical characterization of enzyme candidates can be accomplished using in vitro biochemical assays and/or in vivo bioconversion assays. The in vitro assay entails kinetic studies of the target enzyme on the desirable nonnatural substrate. In the case when a reliable biochemical assay can be established, enzyme kinetics studies provide quantitative comparison between candidate enzymes. However, enzyme purification and kinetics study are time- and resource-consuming. When a large number of candidates are identified from initial database search, in vivo test often serves as a more efficient approach to quick elimination of nonfunctional enzymes. The in vivo bioconversion assay refers to examining the activity of a candidate enzyme in a host strain, which can provide sufficient amount of desirable substrate through either in situ biosynthesis or uptake from media supplement. Meanwhile, the product of the desirable enzyme reaction can either be effectively secreted or be quantitatively converted by subsequent cellular reaction(s) to a secretable metabolite for analysis. Bioconversion assays are also more desirable when a reliable enzyme assay cannot be established due to environmental sensitivity of the enzyme. Furthermore, results from in vivo and in vitro assays are complimentary. Since the unnatural enzyme reaction will be embedded as one step of a biosynthetic pathway in the host, values of k_{cat} and V_{max} help to pinpoint the candidate enzyme's kinetic limitation under cellular condition and enlighten on the future direction of enzyme engineering. On the other hand, the in vivo test can reveal properties, such as protease liability and inhibition by common metabolites, which cannot be easily examined by in vitro studies. Regardless of the characterization method, an ideal enzyme candidate should have good catalytic efficiency, high substrate specificity, low sensitivity to inhibition, easy to express, and reasonable stability inside the host. After a candidate enzyme for the desired unnatural reaction is identified, its orthologs can be screened as the continuation of the enzyme discovery process. The efforts may yield an enzyme with better properties.

3.3 Pathway Engineering

The engineering of a heterologous pathway into the desirable host strain is a multifaceted and iterative process that requires the adjustment of both the pathway enzymes and the metabolic capability of the host strain. With controlling enzyme expression at the center of the problem, pathway optimization is a multivariate task, which requires a systems approach. Such efforts benefit from the collaboration of a multidisciplinary team with knowledge in molecular biology, host genetics, biochemistry, synthetic biology, and systems biology.

3.3.1 Pathway Expression

Unlike a protein purification project, which often targets the maximum gene expression level, installation of a heterologous pathway strikes to achieve the maximum biosynthetic titer, yield, and productivity of the target metabolite in a host strain. This goal requires fine-tuning the expression level of individual enzymes in order to achieve a balance between the kinetic performance of the pathway and metabolic costs associated with protein synthesis. The enzyme expression level can be adjusted at both the transcriptional and the translational levels, as discussed in this chapter.

One major factor that affects the gene dosage is its physical location. By switching from plasmid- to chromosome-based expression, the copy number of an encoding gene can be changed from hundreds to one. At the initial stage of the pathway engineering, different gene combinations often need to be examined. A set of plasmid vectors with different copy number but compatible replication origins and selection markers can be used to simplify the process. Toward the advanced stage of pathway engineering, genes are often integrated into the chromosome to address concerns about the stable maintenance of a plasmid over prolonged time of cultivation. Studies on the chromosome structure of model microbes, such as *E. coli*, reveal that the promoter activity is affected by the superhelical density of the local DNA region (Peter et al. 2004; Dillon and Dorman 2010). Therefore, the chromosomal location of an integrated gene affects its expression level. It was suggested that genes in close proximity to the replication origin of a bacterial chromosome have high expression level due to replication-associated gene dosage effects (Schmid and Roth 1987; Sousa et al. 1997; Rocha 2008), which makes the region attractive for heterologous pathway integration. On the other hand, if the small-molecule product is mainly synthesized at the stationary phase of bacterial growth, in which the replication-associated effect is unclear, choosing a good integration location requires an empirical approach by screening multiple integration sites that are scattered on the chromosome. Integration of the heterologous pathway on bacterial/yeast artificial chromosome or F-factor-derived plasmids is an alternate option for single-copy gene expression. The construction of multi-gene DNA fragment can be greatly facilitated by recently developed DNA assembly methods (Li and Elledge 2007; Gibson et al. 2009).

Modulation of the promoter strength offers another level of control in enzyme expression. Over 40 years (Ippen et al. 1968) of microbial genetics studies have uncovered a wealth of natural promoter elements, which played critical roles in engineering efforts of model organisms. Recent efforts by synthetic biology groups to standardize biological parts yield tools and methodologies for the construction and systematic characterization of synthetic promoter libraries (Solem and Jensen 2002; Alper et al. 2005; Cox et al. 2007; Tyo et al. 2011; Davis et al. 2011). The Registry of Standard Biological Parts hosts a large collection of promoter sequences that can be readily incorporated using BioBrick cloning (Kelly et al. 2009). By altering a natural promoter using error-prone PCR or sequence recombination, promoter elements with varied strength between three to five decades could be created. However, due to the complex nature of the transcription process, a reliable computational algorithm is still unavailable to predict the strength of a particular promoter sequence.

The ribosome binding site (RBS) sequence is central to determining a bacterial mRNA transcript's translation initiation efficiency, which is directly associated with the expression level of the protein product. By applying a thermodynamic model to calculating the free energy change prior and after the 30S complex assembles onto the mRNA, a computer algorithm, RBS calculator, can predict the translation initiation strength of a given RBS sequence (reverse engineering) within a context of 35 bases flanking the start codon. The method can also design RBS sequences with targeted initiation rate (forward engineering) spanning a 100,000+ fold range (Salis et al. 2009). An optimized version with more accurate determination of the 16S rRNA binding site and aligned spacing was later developed (Salis 2011). By combining the function of the RBS calculator and other computational algorithms, the RBS Library Calculator enables the characterization and optimization of a multienzyme biosynthetic pathway through building a sequence-expression-activity map (Farasat et al. 2014).

Riboswitches are untranslated region (UTR) sequences that respond to the intracellular concentrations of specific small-molecule metabolites (ligands). Binding of the aptamer domain of a riboswitch with a ligand induces changes in the secondary structure of a mRNA, which results in the activation or inhibition of the translation (Mandal and Breaker 2004; Dambach and Winkler 2009). Therefore, riboswitches serve as a feasible platform for designing dynamic gene expression circuits that are directly controlled by biosynthetic intermediates (Michener et al. 2012). Versatile and modular riboswitches are successfully designed to demonstrate the capability to execute basic and complex logic functions (Win and Smolke 2007, 2008). For example, two aptamer domains that respectively activate or inhibit the translation by responding to the same ligand are coupled to serve as a bandpass filter in order to maintain relatively stable translation level within a range of the ligand concentrations (Win and Smolke 2008). Application of riboswitch in pathway engineering benefits from the well-established SELEX in vitro selection method in order to modify the sensing domain of the aptamer to a large selection of small molecules (Bunka and Stockley 2006). Progress in designing RNA devices that target other features of mRNA, such as degradation rate, is covered by several recent reviews (Liang et al. 2011; Boyle and Silver 2012; Kang et al. 2014).

3.3.2 Pathway Characterization, Bottleneck Identification and Removal

In the past 20 years, the field of metabolic engineering has built and adapted a plethora of genetic, biochemical, bioanalytical, and computational tools for analyzing different layers in the performance of a biosynthetic pathway. Recent rapid developments in DNA synthesis, DNA/RNA sequencing, mass spectrometry instrumentation, and methodology lead to an omics era that underlines high-throughput technologies as the workhorses for pathway global analysis (Woolston et al. 2013). It provokes the systems-level metabolic engineering approach that focuses on "integrating the 'omic' and computational techniques of systems biology" to take a global-scale snapshot of the biosynthetic process (Kim et al. 2008, 2012; Lee et al. 2012). Although the complete integration of datasets from metabolomics, transcriptomics, fluxomics, and proteomics analysis can be an overwhelming effort (Palsson and Zengler 2010), the proper use of one or the combination of several global analysis methods provides insightful clues to disentangle the complex metabolic network and pinpoint the bottleneck step(s). In general, methods for analyzing and optimizing a natural pathway are also applicable to heterologous pathway engineering. Many excellent books and review articles provide detailed coverage on the basic principle, the methodology, the experimental design, the data analysis, and application examples on these methods (Wittmann and Lee 2012; Alper 2013). In this section, we will focus on aspects that are more relevant in heterologous pathway engineering.

3.3.2.1 Metabolite Profiling

The goal of pathway engineering is to build a biocatalyst for cost-effective production of the target molecule. Therefore, to establish reliable analytical method(s) for the quantification of all intermediates and byproducts that are derived from the pathway is practically the most important step. Instruments such as NMR, GC(-MS), LC(-MS) have all been successfully applied in this process. Due to the trade-off between the coverage on metabolite species and the quality of the data, which refers to the identification and the quantification accuracy, the metabolite profiling of a pathway often focuses on a subset of the total cellular metabolome, especially the secretome, which includes gaseous molecules such as CO_2. To compile a list of key metabolites of a pathway in its native host can benefit from previous biochemical and physiological studies. When a heterologous pathway is installed, however, unexpected metabolic consequences can lead to the formation of byproducts due to unstable biosynthetic intermediates under intracellular condition, promiscuous activities of the pathway enzymes on host metabolites or vice versa. Therefore, monitoring the carbon balance is critical for establishing a complete profile of key metabolites in heterologous pathway engineering. In order to find the missing carbon, heterologous enzymes and nonnative metabolites should be individually examined for its effect on metabolite accumulation. Multiple analytical methods, such as NMR and MS, need to be applied in order to correctly identify the structure of any new metabolite. One example is the engineering of 1,4-butanediol pathway in *E. coli* strain. The formation of γ-butyrolatone from the cyclization of 4-hydroxybutyryl-CoA is due to the instability of the biosynthetic intermediate, which is not part of the pathway design (Yim et al. 2011). The metabolite profiling process also reveals whether a foreign metabolite is cytotoxic, which is often a concern with heterologous pathways.

3.3.2.2 Flux Analysis

Metabolic flux analysis (MFA) reveals the control architecture and the rate at which the biosynthetic pathway operates through a flux distribution diagram that includes a selected set of relevant biochemical reactions (Stephanopouolos et al. 1998; Stephanopoulos 1999; Wiechert 2001). As one major tool in systems metabolic engineering, MFA can also be used to predict the cellular behavior under metabolic perturbations. Several recent reviews and a reference book discussed methodology development and application of MFA (Woolston et al. 2013; Zamboni et al. 2009; Lee et al. 2011; Kromer et al. 2014). Under both genetic and environmental perturbations, the natural metabolic network was shown to be surprisingly robust through rerouting metabolic and catabolic pathways (Ishii et al. 2007). However, as one form of perturbation, the expression of a heterologous pathway can have more profound impacts on the host's metabolism. In the effort to express a novel pathway for the biosynthesis of pyridoxal 5′-phosphate, inhibitory crosstalk mechanisms were discovered between the pathway intermediates and the host's natural enzymes (Kim and Copley 2012).

3.3.2.3 Gene Expression Analysis

Transcriptomics study reveals how a microorganism adjusts to internal and external stimulants at the RNA level. Since the first reports of applying global gene expression profile in microbial research (Wodicka et al. 1997; DeRisi et al. 1997), transcriptome became an indispensable tool in metabolic engineering. Its important role in heterologous pathway engineering is further emphasized due to the hard-to-predict responses of the host's regulatory system to heterologous enzymes and nonnative metabolites. Transcriptomics analysis often provides extremely insightful information on stresses that are imposed by the toxicity of the foreign metabolites, the host's inadequate transport capability, and the energetics or reducing equivalent imbalance caused by the chemical production. Over the years, the technology platform of transcriptomics has evolved from probe-based gene array, which has a definite number of detecting genes, to deep-sequencing-based RNA-seq, which is an open platform that can detect virtually unlimited number of RNA species (Wang et al. 2009). Due to the well-developed sample processing and data analysis protocols, commercial microarrays for model microorganisms are still the workhorses in studies that focus on the gene expression changes of the host. In these cases, one can supplement the analysis with qRT-PCR to obtain the missing data on heterologous genes. On the other hand, RNA-seq is the method of choice when a microarray or even the genome sequence of the host is not available (Gowen and Fong 2010; Shimizu et al. 2013). This platform also overcomes many drawbacks of the hybridization-based methods by providing single nucleotide resolution, better signal-to-noise ratio, and higher dynamic range. As a discovery tool, RNA-seq study of microbial transcription reveals previously unknown or underestimated features that challenge the conventional model (Kim et al. 2012; Güell et al. 2011; Pfeifer-Sancar et al. 2013). This knowledge can potentially lead to novel approaches in metabolic engineering.

3.3.2.4 To Remove Pathway Bottleneck(s)

The first steps to improving the rate of flux through a heterologous pathway often involve the adjustment of pathway enzyme expression levels, using methods discussed in Sect. 3.3.1 of this chapter. Optimizing the spatial organization of pathway enzymes is another effective approach to increase the metabolic productivity (Conrado et al. 2008; Lee et al. 2012). Metabolic channeling is a synergistic effect that is observed in natural multienzyme systems through co-localization in order to reduce the free diffusion and increase the local concentrations of metabolic intermediates. By simply expressing the four subunits of the *E. coli* glucose phosphotransferase system as a single fusion protein, a three to fourfold increase in the catalytic activity was observed (Mao et al. 1995). Enzymes can also be co-localized on protein (Dueber et al. 2009), RNA (Delebecque et al. 2011), or DNA (Conrado et al. 2014) scaffolds to increase the pathway kinetics. The scaffold approach further offers stoichiometric control over enzymes at different steps of a pathway by adjusting the number of the cognate interacting domains on the scaffold. In comparison to modulating the gene expression level, co-localization helps to overcome bottlenecks caused by the inherent kinetic deficiency of pathway enzymes. In addition, by reducing the time gap between production and consumption, unstable biosynthetic intermediates are utilized more efficiently.

Quite often, the kinetic property of the pathway enzyme(s), especially the one that operates on a nonnative substrate, is the limiting step in an artificially assembled heterologous pathway. Enzyme engineering is therefore required to generate mutant enzymes with desirable properties. An enzyme evolution experiment consists of iterative rounds of sequence diversification, mutant library selection/screening, and hit characterization, until a desirable mutant is obtained or an evolutionary plateau is reached. The same set of criteria that are used to select the best candidate pathway enzyme (see Sect. 3.2.3) is applicable to this process. A key step in the enzyme engineering process is to establish a reliable high-throughput library selection/screening method, which often dictates the choice of sequence diversification method (Arnold and Georgiou 2003a, b; Reymond 2006; Gillam et al. 2014). In a selection method, the catalytic activity of the target enzyme is associated with certain phenotype of the host strain. The experiment involves transforming the mutant library into a selection host strain, plating, and culturing the transformants on agar plates under selective pressure (Shen et al. 2011). Since a selection method can simultaneously process hundreds of millions of mutants, its throughput is often limited by the diversity of the mutant library. However, a large number of false positive clones may yield from a selection method due to microbes' adaptability to selection pressure. Furthermore, it is not always possible to link enzyme activity to an easily observable phenotype. A screening method refers to examining mutant enzymes expressed by individual clone using in vitro biochemical assays, in vivo bioconversion assays, or phenotype sorting. Application of such method is severely constrained by the availability of high-throughput instrumentation, such as colony picking and liquid transfer robots, or a flow cytometer if a Fluorescence-Activated Cell Sorting (FACS) method is desirable. Regardless of the approach used, building a direct link between the signal readout and the enzyme activity is the key to minimize the frequency of false positives/negatives. Enzyme engineering efforts can be greatly assisted when the structure of the parent enzyme is available. By docking the nonnative substrate molecule into the active site of the parent enzyme using tools, such as AUTODOCK (Morris et al. 2009) or GOLD (Verdonk et al. 2003), residues that contribute to binding energy increase can be identified for mutagenesis. As an improvement from above method, the Combinatorial Active-site Saturation Test (CAST) systematically explores the contribution of active site residues to the binding of the nonnative substrate through iterative round of saturation mutagenesis (Reetz et al. 2005, 2006; Reetz 2011). In each round, a couple of chosen residues are randomized to form a focused library, which is screened for mutants with desired property. The best hit obtained in the current round is used as the template for the next round of diversification at other chosen sites. The method effectively minimizes the library size, reduces the screening effort to low or medium throughput, while still virtually covers all the sequence space. CAST was demonstrated as a more effective method than traditional ones in an attempt to improve the stereoselectivity of a lipase (Reetz et al. 2010).

3.3.2.5 Host Strain Engineering

Step(s) that confine the performance of a heterologous pathway can reside in the metabolic capability of the host as well as in the kinetic performance of the pathway. Host strain engineering often requires genetic manipulations, such as deletion of native gene, insertion of foreign gene cassette, and replacement of regulatory region, in order to eliminate competing pathways, minimize byproduct formation, increase cofactor availability, and improve transporter function. Molecular biology methods that target single-site genetic changes are well developed for model microorganisms (Datsenko and Wanner 2000; Sharan et al. 2009). The majority of the methods rely on the host's (e.g., yeast) or a phage-derived (λ Red system) homologous recombination machinery. The CRISPR-Cas9 system provides an alternate method for precision cleavage in the host's genome sequence (Doudna and Charpentier 2014; Jakočiūnas et al. 2016). Various genetic markers were exploited for the efficient identification of successful crossover events through phenotypical selection. Recently developed Multiplexed Automated Genome Engineering (MAGE) method, which relies on single-strand DNA-mediated allelic exchange, drastically improves the throughput of the process through the introduction of multiple targeted changes on the chromosome at the same time. In conjunction with hierarchical conjugative assembly to minimize the deleterious effect of the required *mutS* deletion, the MAGE method was applied to the replacement of 314 TAG stop codons with TAA codons on *E. coli* chromosome (Isaacs et al. 2011). Besides generating defined changes, MAGE was also used in a combinatorial approach to achieve genome-wide strain engineering. By simultaneously randomizing the RBS sequences of 20 genes that modulate the 1-deoxy-D-xylulose-5-phosphate (DXP) pathway in *E. coli*, a strain with fivefold increase in lycopene production was identified (Wang et al. 2009). This approach is a major deviation from conventional strain engineering methods, which focus on examining the metabolic effect one gene at a time. The genome-wide approach is especially advantageous when multiple factors need to be addressed to improve the metabolic capability of the host. Through simultaneously exploring multiple variables, the strain engineering process can be rapidly accelerated. Other genome-wide strain engineering methods include the trackable multiplex recombineering (TRMR) (Warner et al. 2010) and global transcription machinery engineering (gTME) (Alper et al. 2006). As an approach with a broad range of targets, gTME method is effective on improving the product tolerance of the yeast host, which has been proven to be a difficult trait to engineer.

3.4 Heterologous Synthesis of 1-Butanol as an Example

1-Butanol is an important industrial platform molecule and a potential fuel substitute. It is accumulated by *Clostridia* strains as a natural metabolite at solventogenic growth phase. Due to challenges in genetic manipulations and complex metabolic characteristics of its natural producers, heterologous production of 1-butanol has attracted significant research efforts in recent years (Kumar and Gayen 2011; Jang et al. 2012; Lan and Liao 2013). Microbial hosts, including *E. coli*, *Saccharomyces cerevisiae*, *Bacillus subtilis*, *Pseudomonas putida*, *Bacillus subtilis*, and *Lactobacillus brevis*, have been examined. Another reason to explore alternate production hosts lies in the toxicity of 1-butanol, which inhibits the growth of many microorganisms at a concentration of 2 % (w/v) (Knoshaug and Zhang 2009).

3.4.1 *Engineering Modified Natural 1-Butanol Pathway in* E. coli

The heterologous production of 1-butanol was first demonstrated in *E. coli* host by Atsumi et al. (2008). Genes encoding the *Clostridium acetobutylicum* 1-butanol biosynthetic pathway were amplified and expressed in JCL16 (Fig. 3.2a). Under anaerobic growth condition, the strain produced 13.6 mg/L of 1-butanol. Replacing the *C. acetobutylicum* acetyl-CoA acetyltransferase with *E. coli* AtoB enzyme led to more than threefold increase in 1-butanol accumulation, presumably because of higher activity of AtoB (Atsumi et al. 2008). Shen et al. further focused on overcoming the reversibility and improving the robustness of the natural pathway to optimize the biosynthesis in *E. coli* (Fig. 3.2) (Shen et al. 2011). The reduction of crotonyl-CoA to butyryl-CoA is a key step in 1-butanol pathway. Although the *Clostridia* enzyme Bcd, together with its redox partner EftAB, did support the production of 1-butanol, no butyryl-CoA dehydrogenase activity was detected in the cell lysate. Taken the discrepancy between the in vivo and in vitro analysis as the result of lacking robustness in the Bcd system, several enzyme candidates that can catalyze the reduction of crotonyl-CoA were screened for higher 1-butanol production. The *trans*-enoyl-CoA reductase (Ter) from *Treponema denticola* led to the highest level of 1-butanol accumulation (1.8 g/L). Since the Ter enzyme does

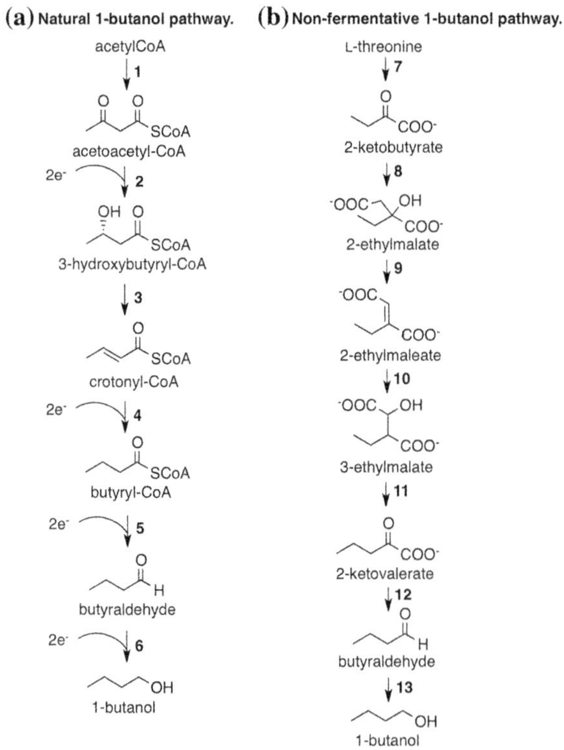

Fig. 3.2 1-Butanol biosynthetic pathways (Lan and Liao 2013; Shen et al. 2011). Enzymes (encoding genes): **1** acetoacetyl-CoA thiolase (*thl*); **2** 3-hydroxybutyryl-CoA dehydrogenase (*hbd*); **3** crotonase (*crt*); **4** butyryl-CoA dehydrogenase (*bcd*) and electron transfer flavoprotein (*etf*); **5** and **6** aldehyde/alcohol dehydrogenase (*adhE2*); **7** threonine deaminase (*ilvA*); **8** 2-ethylmalate synthase (*leuA*); **9** and **10** 2-ethylmalate isomerase (*leuCD*); **11** 3-ethylmalate dehydrogenase (*leuB*); **12** 2-keto acid decarboxylase (*kivd*); **13** alcohol dehydrogenase (*adh2*)

not catalyze the oxidation of butyroyl-CoA, it serves as a check valve to prevent metabolic backflux. Host strain metabolic engineering strategies were then applied to increasing the intracellular availability of acetyl-CoA and NADH in order to create a "driving force" to shuttle the carbon flux through the 1-butanol pathway. The combined manipulation led to a strain that accumulated 15 g/L of 1-butanol under anaerobic condition at 88 % of the theoretical yield in 3 days. A similar strategy to improve the pathway kinetics was used by Bond-Watts et al. (2011). They independently identified Ter as an irreversible enzyme both in vitro and in vivo. Instead of improving the anaerobic NADH availability, the pyruvate dehydrogenase complex was overexpressed. The resulting strain produced an average of 4.65 g/L 1-butanol under aerobic condition at 28 % yield from glucose. Shen et al. introduced two process engineering methods to further improve the biosynthesis. They implemented a two-stage fermentation process, which constitutes an aerobic growth phase and an anaerobic production phase. Coupled with continuous gas stripping for in situ 1-butanol removal, approximately 30 g/L of 1-butanol was produced at 70 % of the theoretical yield in seven days. It currently represents the highest titer achieved by a heterologous host (Shen et al. 2011).

3.4.2 Engineering Artificial 1-Butanol Pathways in E. coli

To overcome inherent problems of the natural 1-butanol biosynthetic pathway, several artificial routes were designed. Atsumi et al. first described a non-fermentative pathway (Fig. 3.2b) that is based on the Ehrlich pathway of 2-keto acid degradation, which is a virtually irreversible process under normal microbial growth condition (Atsumi et al. 2008). The artificial pathway consists multiple steps that require enzyme activities on nonnative substrates, including the conversion of 2-ketobutyrate, which is derived from the deamination of L-threonine, into key intermediate 2-ketovalerate, and the further decarboxylation of 2-ketovalerate. To complete the first conversion, the promiscuous *E. coli* LeuABCD enzymes, which function in L-leucine biosynthesis, were used. After screening five 2-keto acid decarboxylases with varied substrate specificity in an in vivo bioconversion assay, the *kivd*-encoded α-ketoisovalerate decarboxylase from *Lactococcus lactis* was identified as the most active one on 2-ketovalerate. An *E. coli* strain overexpressing all pathway enzymes synthesized 3.2 mM 1-butanol from 8 g/L of L-threonine that was supplemented in the culture media. Using a host strain with elevated metabolic flux into L-threonine pathway, de novo biosynthesis of 1-butanol from glucose was achieved at around 1 g/L (Shen and Liao 2008). Due to the side-activity of Kivd on 2-ketobutyrate, the strain also co-produced 1-propanol. In a modified version of the non-fermentative pathway, citramalate synthase from *Methanococcus jannaschii* was installed to create an alternate and shorter route to 2-ketobutyrate through the condensation between acetyl-CoA and pyruvate (Ataumi and Liao 2008). To improve the activity of the thermophilic citramalate synthase at ambient temperature, a mutant library was generated by error-prone PCR and selected using a growth-based scheme. An *E. coli* strain expressing the alternate route synthesized 0.5 g/L 1-butanol in 92 h. Although the titer and yield achieved by the non-fermentative pathway are lower than those of the modified natural pathway, further optimization, especially enzyme engineering, can lead to significant improvement. In a separate report, Zhang et al. identified a LeuA G462D mutant, which led to over 27-fold improvement in 1-butanol accumulation (Zhang et al. 2008). Furthermore, structure-based mutagenesis was proven to be successful in engineering substrate specificity of Kivd (Zhang et al. 2008). By using a Kivd variant that can better differentiate 2-ketovalerate from 2-ketobutyrate, the titer of 1-butanol synthesis can be improved through reducing the 1-propanol formation.

One unique example of artificial 1-butanol biosynthetic pathway engineering exploited the reversal of the β-oxidation cycle, which follows the same reaction sequence as the natural pathway, but completely relies on activities of *E. coli* native enzymes (Dellomonaco et al. 2011). Dellomonaco et al. created an *E. coli* host with constitutive expression of the fatty acid β-oxidation pathway and deletions in several fermentation pathways. Overexpression of enzymes that catalyze the first step, acetyl-CoA condensation, and the last step, butyraldehyde reduction, of the 1-butanol pathway led to the accumulation of 1.9 g/L 1-butanol in 24 h. Following additional chromosomal deletions to reduce byproduct formation, the new strain produced 14 g/L 1-butanol within 48 h under batch fermentation condition. Subsequent analysis revealed the identities of *E. coli* enzymes that catalyzed each step of the biosynthesis. This approach benefits from the versatile synthetic power of *E. coli* native enzymes.

3.4.3 Other Lessons

Besides successes in titer and yield optimization of 1-butanol biosynthesis, strategies and tools that are developed in the process can benefit engineering efforts in other heterologous pathways. Dietrich et al. designed a transcription factor-based biosensor for the screening/selection of *E. coli* strains with improved 1-butanol synthesis (Dietrich et al. 2013). The sensing mechanism was based on the σ^{54}-transcription activator, BmoR, and the σ^{54}-dependent, alcohol-regulated promoter, P_{BMO} from *Thauera butanivorans*. Upon P_{BMO}-initiated transcription of a reporter gene, the *E. coli* host displayed phenotype changes that correlate with the concentrations of externally added 1-butanol. The strategy was successfully applied to the identification of *E. coli* variants with 35 % increase in 1-butanol specific productivity. In the process of optimizing the Ter enzyme activity for 1-butanol synthesis, Shen et al. developed a growth-coupled strategy to detect Ter variants with higher intracellular activity (Shen et al. 2011). The method used an *E. coli* host that lost anaerobic growth due to NADH surplus caused by deletions of multiple fermentation pathways. Since Ter is a NADH-dependent enzyme, a mutant with higher catalytic activity can quickly regenerate NAD^+ and results in a faster growth phenotype under anaerobic condition. Application of this method led to the identification of a Ter mutant that resulted in fivefold increase in 1-butanol production. As a potentially general selection scheme, the method can be used in directed evolution of NADH-dependent enzymes.

References

Alper HS (ed) (2013) Systems metabolic engineering: Methods and protocols. In: Methods in molecular biology. Springer, New York

Alper H, Fischer C, Nevoigt E, Stephanopoulos G (2005) Tuning genetic control through promoter engineering. Proc Natl Acad Sci USA 102:12678–12683

Alper H, Moxley J, Nevoigt E, Fink GR, Stephanopoulos G (2006) Engineering yeast transcription machinery for improved ethanol tolerance and production. Science 314:1565–1568

Andersson SG, Kurland CG (1990) Codon preferences in free-living microorganisms. Microbiol Rev 54:198–210

Arnold FH (2001) Combinatorial and computational challenges for biocatalyst design. Nature 409:253–257

Arnold FH, Georgiou G (eds) (2003a) Directed enzyme evolution: screening and selection methods. In: Methods in molecular biology. Springer, New York

Arnold FH, Georgiou G (eds) (2003b) Directed evolution library creation: methods and protocols. In: Methods in molecular biology. Springer, New York

Ataumi S, Liao JC (2008) Directed evolution of *Methanococcus jannaschii* citramalate synthase for biosynthesis of 1-propanol and 1-butanol by *Escherichia coli*. Appl Environ Microbiol 74:7802–7808

Atsumi S, Cann AF, Connor MR, Shen CR, Smith KM, Brynildsen MP, Chou KJ, Hanai T, Liao JC (2008a) Metabolic engineering of *Escherichia coli* for 1-butanol production. Metab Eng 10:305–311

Atsumi S, Hanai T, Liao JC (2008b) Non-fermentative pathways for synthesis of branched-chain higher alcohols as biofuels. Nature 451:86–89

Bond-Watts BB, Bellerose RJ, Chang MC (2011) Enzyme mechanism as a kinetic control element for designing synthetic biofuel pathways. Nat Chem Biol 7:222–227

Bounton ZL, Bennet GN, Rudolph FB (1996) Cloning, sequencing, and expression of clustered genes encoding beta-hydroxybutyryl-coenzyme A (CoA) dehydrogenase, crotonase, and butyryl-CoA dehydrogenase from *Clostridium acetobutylicum* ATCC 824. J Bacteriol 178:3015–3024

Boyle PM, Silver PA (2012) Parts plus pipes: synthetic biology approaches to metabolic engineering. Metab Eng 14:223–232

Bunka DH, Stockley PG (2006) Aptamers come of age—at last. Nat Rev Microbiol 4:588–596

Conrado RJ, Varner JD, DeLisa MP (2008) Engineering the spatial organization of metabolic enzymes: mimicking nature's synergy. Curr Opin Biotechnol 19:492–499

Conrado RJ, Wu GC, Boock JT, Xu H, Chen SY, Lebar T, Turnsek J, Tomsic N, Avbeij M, Gaber R, Koprivnjak T, Mori J, Glavnik V, Vovk I, Bencina M, Hodnik V, Anderluh G, Dueber JE, Jerala R, DeLisa DP (2014) DNA-guided assembly of biosynthetic pathways promotes improved catalytic efficiency. Nucleic Acids Res 40:1879–1889

Corey EJ (1988) Retrosynthetic thinking-essentials and examples. Chem Soc Rev 17:111–133

Cox RSIII, Surette MG, Elowitz MB (2007) Programming gene expression with combinatorial promoters. Mol Syst Biol 3:145

Dambach MD, Winkler WC (2009) Expanding roles for metabolite-sensing regulatory RNAs. Curr Opin Microbiol 12:161–169

Daniel R (2002) Construction of environmental libraries for functional screening of enzyme activity. In: Brakmann S, Johnsson K (eds) Directed molecular evolution of proteins, Wiley-VCH,Weinheim

Datsenko KA, Wanner BL (2000) One-step inactivation of chromosomal genes in Escherichia coli K-12 using PCR products. Proc Natl Acad Sci USA 97:6640–6645

Davis JH, Rubin AJ, Sauer RT (2011) Design, construction and characterization of a set of insulated bacterial promoters. Nucleic Acids Res 39:1131–1141

Delebecque CJ, Lindner AB, Silver PA, Aldaye FA (2011) Organization of intracellular reactions with rationally designed RNA assemblies. Science 333:470–474

Dellomonaco C, Clomburg JM, Miller EN, Gonzalez R (2011) Engineered reversal of the β-oxidation cycle for the synthesis of fuels and chemicals. Nature 476:355–359

DeLong EF (ed) (2013) Microbial metagenomics, metatranscriptomics, and metaproteomics. Methods Enzymology. Academic Press, San Diego

DeRisi JL, Iyer VR, Brown PO (1997) Exploring the metabolic and genetic control of gene expression on a genomic scale. Science 278:680–686

Dietrich JA, Shis DL, Alikhani A, Keasling JD (2013) Transcription factor-based screens and synthetic selections for microbial small-molecule biosynthesis. ACS Synth Biol 2:47–58

Dillon SC, Dorman CJ (2010) Bacterial nucleoid-associated proteins, nucleoid structure and gene expression. Nat Rev Microbiol 8:185–195

Doudna JA, Charpentier E (2014) The new frontier of genome engineering with CRISPR-Cas9. Science 346:1258096

Draths KM, Frost JW (1994) Environmentally compatible synthesis of adipic acid from D-glucose. J Am Chem Soc 116:399–400

Dueber JE, Wu GC, Malmirchegini GR, Moon TS, Petzold CJ, Ullal AV, Prather KL, Keasling JD (2009) Synthetic protein scaffolds provide modular control over metabolic flux. Nat Biotechnol 27:753–759

Farasat I, Kushwaha M, Collens J, Easterbrook M, Guido M, Salis HM (2014) Efficient search, mapping, and optimization of multi-protein genetic systems in diverse bacteria. Mol Syst Biol 10:731

Gibson DG, Yong L, Chuang RY, Venter JC, Hutchison CA 3rd, Smith HO (2009) Enzymatic assembly of DNA molecules up to several hundred kilobases. Nat Methods 6:343–345

Gillam EMJ, Copp JN, Ackerley DF (eds) (2014) Directed evolution library creation: methods and protocols, 2nd edn. In: Methods in molecular biology. Springer, New York

Gowen CM, Fong SS (2010) Genome-scale metabolic model integrated with RNAseq data to identify metabolic states of Clostridium thermocellum. Biotechnol J 5:759–767

Güell M, Yus E, Lluch-Senar M, Serrano L (2011) Bacterial transcriptomics: what is beyond the RNA horiz-ome? Nat Rev Microbiol 12:658–669

Gustafsson C, Govindarajan S, Minshull J (2004) Codon bias and heterologous protein expression. Trends Biotechnol 22:346–353

Gutman GA, Hatfield GW (1989) Nonrandom utilization of codon pairs in *Escherichia coli*. Proc Natl Acad Sci USA 86:3699–3703

Handelsman J, Rondon MR, Brady SF, Clardy J, Goodman RM (1998) Molecular biological access to the chemistry of unknown soil microbes: a new frontier for natural products. Chem Biol 5:R245–R249

Ippen K, Miller JH, Scaife J, Beckwith J (1968) New controlling element in the *Lac* operon of *E. coli*. Nature 217:825–827

Isaacs FJ, Carr PA, Wang HH, Lajoie MJ, Sterling B, Kraal L, Tolonen AC, Gianoulis TA, Goodman DB, Reppas NB, Emig CJ, Bang D, Hwang SJ, Jewett MC, Jacobson JM, Church GM (2011) Precise manipulation of chromosomes in vivo enables genome-wide codon replacement. Science 333:348–353

Ishii N et al (2007) Multiple high-throughput analyses monitor the response of *E. coli* to perturbations. Science 316:593–597

Jakočiūnas T, Jensen MK, Keasling JD (2016) CRISPR/Cas9 advances engineering of microbial cell factories. Metab Eng 34:44–59

Jang YS, Lee J, Malaviya A, Seung DY, Cho JH, Lee SY (2012) Butanol production from renewable biomass: rediscovery of metabolic pathways and metabolic engineering. Biotechnol J 7:186–198

Kang Z, Zhang C, Zhang J, Jin P, Zhang J, Du G, Chen J (2014) Small RNA regulators in bacteria: powerful tools for metabolic engineering and synthetic biology. Appl Microbiol Biotechnol 98:3413–3424

Kelly JR, Rubin AJ, Davis JH, Ajo-Franklin CM, Cumbers J, Czar MJ, de Mora K, Glieberman AL, Monie DD, Endy D (2009) Measuring the activity of BioBrick promoters using an in vivo reference standard. J Biol Eng. 3, doi:10.1186/1754-1611-3-4

Kim J, Copley SD (2012) Inhibitory cross-talk upon introduction of a new metabolic pathway into an existing metabolic network. Proc Natl Acad Sci USA 109:E2856–E2864

Kim TY, Sohn SB, Kim HU, Lee SY (2008) Strategies for systems-level metabolic engineering. Biotechnol J 3:612–623

Kim IK, Roldão A, Siewers V, Nielsen J (2012) A systems-level approach for metabolic engineering of yeast cell factories. FEMS Yeast Res 12:228–248

Knoshaug EP, Zhang M (2009) Butanol tolerance in a selection of microorganisms. Appl Biochem Biotechnol 153:13–20

Kromer JO, Nielsen L, Blank LM (eds) (2014) Metabolic flux analysis: methods and protocols. In: Methods in molecular biology. Springer, New York

Kumar M, Gayen K (2011) Developments in biobutanol production: new insights. Appl Energy 88:1999–2012

Lan EI, Liao JC (2013) Microbial synthesis of n-butanol, isobutanol, and other higher alcohols from diverse resources. Bioresour Technol 135:339–349

Lee SY, Park JM, Kim TY (2011) Application of metabolic flux analysis in metabolic engineering. Methods Enzymol 498:67–93

Lee JW, Na D, Park JM, Lee J, Choi S, Lee SY (2012a) Systems metabolic engineering of microorganisms for natural and non-natural chemicals. Nat Chem Biol 17:536–546

Lee H, DeLoache WC, Dueber JE (2012b) Spatial organization of enzymes for metabolic engineering. Metab Eng 14:242–251

Li MZ, Elledge SJ (2007) Harnessing homologous recombination in vitro to generate recombinant DNA via SLIC. Nat Methods 4:251–256

Liang JC, Bloom RJ, Smolke CD (2011) Engineering biological systems with synthetic RNA molecules. Mol Cell 43:915–926

Mandal M, Breaker RR (2004) Gene regulation by riboswitches. Nat Rev Mol Cell Biol 5:451–463

Mao Q, Schunk T, Gerber B, Erni B (1995) A string of enzymes, purification and characterization of a fusion protein comprising the four subunits of the glucose phosphotransferase system of *Escherichia coli*. J Biol Chem 270:18295–18300

Medema MH, van Raaphorst R, Takano E, Breitling R (2012) Computational tools for the synthetic design of biochemical pathways. Nat Rev Microbiol 10:191–202

Michener JK, Thodey K, Liang JC, Smolke CD (2012) Applications of genetically-encoded biosensors for the construction and control of biosynthetic pathways. Metab Eng 14:212–222

Minshull J, Stemmer WP (1999) Protein evolution by molecular breeding. Curr Opin Chem Biol 3:284–290

Mori K, Tobimatsu T, Hara T, Toraya T (1997) Characterization, sequencing, and expression of the genes encoding a reactivating factor for glycerol-inactivated adenosylcobalamin-dependent diol dehydratase. J Biol Chem 272:32034–32041

Morris GM, Huey R, Lindstrom W, Sanner MF, Belew RK, Goodsell DS, Olson AJ (2009) AutoDock4 and AutoDockTools4: automated docking with selective receptor flexibility. J Comput Chem 30:2785–2791

Nakamura CE, Whited GM (2003) Metabolic engineering for the microbial production of 1,3-propanediol. Curr Opin Biotechnol 14:454–459

Nakamura Y, Gojobori T, Ikemura T (2000) Codon usage tabulated from the international DNA sequence databases: status for the year 2000. Nucleic Acids Res 28:292

Nassal M, Mogi T, Karnik SS, Khorana HG (1987) Structure-function studies on bacteriorhodopsin III. Total synthesis of a gene for bacterio-opsin and its expression in *Escherichia coli*. J Biol Chem 262:9264–9270
Niu W, Molefe MN, Frost JW (2003) Microbial synthesis of the energetic material precursor 1,2,4-butanetriol. J Am Chem Soc 125:12998–12999
Novoa EM, de Pouplans RL (2012) Speeding with control: codon usage, tRNAs, and ribosomes. Trends Genet 28:574–581
Paddon CJ, Keasling JD (2014) Semi-synthetic artemisinin: a model for the use of synthetic biology in pharmaceutical development. Nat Rev Microbiol 12:355–367
Paddon CJ, Westfall PJ, Pitera DJ, Benjamin K, Fisher K, McPhee D, Leavell MD, Tai A, Main A, Eng D, Polichuk DR, Teoh KH, Reed DW, Treynor T, Lenihan J, Fleck M, Bajad S, Dang G, Dengrove D, Diola D, Dorin G, Ellens KW, Fickes S, Galazzo J, Gaucher SP, Geistlinger T, Henry R, Hepp M, Horning T, Iqbal T, Jiang H, Kizer L, Lieu B, Melis D, Moss N, Regentin R, Secrest S, Tsuruta H, Vazquez R, Westblade LF, Xu L, Yu M, Zhang Y, Zhao L, Lievense J, Covello PS, Keasling JD, Reiling KK, Renninger NS, Newman JD (2013) High-level semi-synthetic production of the potent antimalarial artemisinin. Nature 496: 528–532
Palsson B, Zengler K (2010) The challenge of integrating multi-omic data sets. Nat Chem Biol 6:787–789
Peter BJ, Arsuaga J, Breier AM, Khodursky AB, Brown PO, Cozzarelli NR (2004) Genomic transcriptional response to loss of chromosomal supercoiling in *Escherichia coli*. Genome Biol 5:R87
Pfeifer-Sancar K, Mentz A, Rüchert C, Kalinowski J (2013) Comprehensive analysis of the *Corynebacterium glutamicum* transcriptome using an improved RNAseq technique. BMC Genom 14:888
Plotkin JB, Kudla G (2011) Synonymous but not the same: the causes and consequences of codon bias. Nat Rev Genet 12:32–42
Quadri LE, Weinreb PH, Lei M, Nakano MM, Zuber P, Walsh CT (1998) Characterization of Sfp, a *Bacillus subtilis* phosphopantetheinyl transferase for peptidyl carrier protein domains in peptide synthetases. Biochemistry 37:1585–1595
Rahman SA, Cuesta SM, Furnham N, Holliday GL, Thornton JM (2014) EC-BLAST: a tool to automatically search and compare enzyme reactions. Nat Methods 11:171–174
Reetz MT (2011) Laboratory evolution of stereoselective enzymes: a prolific source of catalysts for asymmetric reactions. Angew Chem Int Ed Engl 50:138–174
Reetz MT, Bocola M, Carballeira JD, Zha D, Vogel A (2005) Expanding the range of substrate acceptance of enzymes: combinatorial active-site saturation test. Angew Chem Int Ed Engl 44:4192–4196
Reetz MT, Wang LW, Bocola M (2006) Directed evolution of enantioselective enzymes: iterative cycles of CASTing for probing protein-sequence space. Angew Chem Int Ed Engl 45: 1236–1241
Reetz MT, Prasad S, Carballeira JD, Gumulya Y, Bocola M (2010) Iterative saturation mutagenesis accelerates laboratory evolution of enzyme stereoselectivity: rigorous comparison with traditional methods. J Am Chem Soc 132:9144–9152
Reymond J-L (ed) (2006) Enzyme assays: high-throughput screening, genetic selection and fingerprinting. Wiley-VCH,Weinheim
Richter F, Leaver-Fay A, Khare SD, Bjelic S, Baker D (2011) De novo enzyme design using Rosetta3. PLoS One 6:e19230
Ro DK, Paradise EM, Ouellet M, Fisher KJ, Newman KL, Ndungu JM, Ho KA, Eachus RA, Ham TS, Kirby J, Chang MC, Withers ST, Shiba Y, Sarpong R, Keasling JD (2006) Production of the antimalarial drug precursor artemisinic acid in engineered yeast. Nature 440:940–943
Rocha EP (2008) The organization of the bacterial genome. Annu Rev Genet 42:211–233

Röthlisberger D, Khersonsky O, Wollacott AM, Jiang L, DeChancie J, Betker J, Gallaher JL, Althoff EA, Zanghellini A, Dym O, Albeck S, Houk KN, Tawfik DS, Baker D (2008) Kemp elimination catalysts by computational enzyme design. Nature 453:190–195

Salis HM (2011) The ribosome binding site calculator. Methods Enzymol 498:19–42

Salis HM, Mirsky EA, Voigt CA (2009) Automated design of synthetic ribosome binding sites to control protein expression. Nat Biotechnol 27:946–950

Schmid MB, Roth JR (1987) Gene location affects expression level in *Salmonella typhimurium*. J Bacteriol 169:2872–2875

Schmidt-Dannert C (2001) Directed evolution of single proteins, metabolic pathways, and viruses. Biochemistry 40:13125–13136

Sharan SK, Thomason LC, Kunzetsove SG, Court DL (2009) Recombineering: a homologous recombination-based method of genetic engineering. Nat Protocol 4:206–223

Shen CR, Liao JC (2008) Metabolic engineering of *Escherichia coli* for 1-butanol and 1-propanol production via the keto-acid pathways. Metab Eng 10:312–320

Shen CR, Lan EI, Dekishima Y, Baez A, Cho KM, Liao JC (2011) Driving forces enable high-titer anaerobic 1-butanol synthesis in *Escherichia coli*. Appl Environ Microbiol 11:2905–2915

Shimizu R, Chou K, Orita I, Suzuki Y, Nakamura S, Fukui T (2013) Detection of phase-dependent transcriptomic changes and Rubisco-mediated CO_2 fixation into poly (3-hydroxybutyrate) under heterotrophic condition in *Ralstonia eutropha* H16 based on RNA-seq and gene deletion analyses. BMC Microbiol 13:169

Siegel JB, Zanghellini A, Lovick HM, Kiss G, Lambert AR, St Clair JL, Gallaher JL, Hilvert D, Gelb MH, Stoddard BL, Houk KN, Michael FE, Baker D (2010) Computational design of an enzyme catalyst for a stereoselective bimolecular Diels-Alder reaction. Science. 329, 309–313

Simon C, Daniel R (2011) Metagenomic analysis: Past and future trends. Appl Environ Microbiol 77:1153–1161

Solem C, Jensen PR (2002) Modulation of gene expression made easy. Appl Environ Microbiol 68:2397–2403

Sousa C, de Lorenzo V, Cebolla A (1997) Modulation of gene expression through chromosomal positioning in *Escherichia coli*. Microbiology 143:2071–2078

Stephanopoulos G (1999) Metabolic fluxes and metabolic engineering. Metab Eng 1:1–11

Stephanopouolos GN, Aristidou AA, Nielsen J (1998) Metabolic engineering: principles and methodologies. Academic Press, San Diego

Tyo KE, Nevoigt E, Stephanopoulos G (2011) Directed evolution of promoters and tandem gene arrays for customizing RNA synthesis rates and regulation. Methods Enzymol 497:135–155

Verdonk ML, Cole JC, Hartshorn MJ, Murray CW, Taylor RD (2003) Improved protein-ligand docking using GOLD. Proteins 52:609–623

Wang Z, Gerstein M, Snyder M (2009a) RNA-Seq: a revolutionary tool for transcriptomics. Nat Rev Genet 10:57–63

Wang HH, Isaacs FJ, Carr PA, Sun ZZ, Xu G, Forest CR, Church GM (2009b) Programming cells by multiplex genome engineering and accelerated evolution. Nature 460:894–898

Warner JR, Reeder PJ, Karimpour-Fard A, Woodruff LB, Gill RT (2010) Rapid profiling of a microbial genome using mixtures of barcoded oligonucleotides. Nat Biotechnol 28:856–862

Welch M, Villalobos A, Gustafsson C, Minshull J (2011) Designing genes for successful protein expression. Methods Enzymol 498:43–66

Wiechert W (2001) ^{13}C metabolic flux analysis. Metab Eng 3:195–206

Win MN, Smolke CD (2007) A modular and extensible RNA-based gene-regulatory platform for engineering cellular function. Proc Natl Acad Sci USA 104:14283–14288

Win MN, Smolke CD (2008) Higher-order cellular information processing with synthetic RNA devices. Science 322:456–460

Wittmann C, Lee SY (eds) (2012) Systems metabolic engineering. Springer, New York

Wodicka L, Dong H, Mittmann M, Ho M-H, Lockhart DJ (1997) Genome-wide expression monitoring in *Saccharomyces cerevisiae*. Nat Biotechnol 15:1359–1367

Woolston BM, Edgar S, Stephanopouos G (2013) Metabolic engineering: past and future. Annu Rev Chem Biomol Eng 4:259–288

Yim H, Haselbeck R, Niu W, Pujol-Baxley C, Burgard A, Boldt J, Khandurina J, Trawick JD, Osterhout RE, Stephen R, Estadila J, Teisan S, Schreyer B, Andrae S, Yang TH, Lee SY, Burk MJ, Van Dien S (2011) Metabolic engineering of *Escherichia coli* for direct production of 1,4-butanediol. Nat Chem Biol 7:445–452

Zamboni N, Fendt SM, Rühl M, Sauer U (2009) ^{13}C-based metabolic flux analysis. Nat Protoc 4:878–892

Zhang K, Sawaya MR, Eisenberg DS, Liao JC (2008) Expanding metabolism for biosynthesis of nonnatural alcohols. Proc Natl Acad Sci USA 105:20653–20658

Chapter 4
Leveraging Gene Synthesis, Advanced Cloning Techniques, and Machine Learning for Metabolic Pathway Engineering

Kedar G. Patel, Mark Welch and Claes Gustafsson

Abstract Modulation and optimization of metabolic pathways is accomplished by several complementary approaches that influence the presence, catalytic properties, and abundance of pathway enzymes. System-wide approaches can also provide an alternative means to influence pathway performance. Current gene synthesis technologies and other molecular tools enable the manipulation of biological systems at the individual component or part level as well as from a whole genome perspective. Our ability to precisely engineer large DNA sequences has matured over the past few decades to enable facile de novo synthesis of genes, vectors, pathways, and even entire chromosomes with any desired nucleotide sequence. We are no longer confined to the cloning and limited manipulation of naturally occurring DNA sequences to engineer or transplant pathways for the production of natural or novel compounds in a favorable host organism. With gene synthesis technologies, DNA parts and assemblies with virtually any imaginable DNA sequence can be created and introduced into any production host system for metabolic pathway. Biological diversity space is vast. In comparison, our ability to navigate this multidimensional space is limited. Reliably navigating this mega-dimensional space requires versatile control over DNA sequence. DNA2.0 has developed a bioengineering platform that seamlessly integrates gene synthesis, genome editing, and modern machine learning for whole bio-system optimization. This approach enables exploration of a large number of variables (e.g., synonymous mutations, amino acid substitutions, DNA or protein parts all the way to pathway replacement and genome-level modifications) while minimizing the number of samples needed. A key to our approach is the broad, unbiased sampling of targeted sequence variables. Causal variables are

K.G. Patel · M. Welch · C. Gustafsson (✉)
DNA2.0 Inc., 37950 Central Court, Newark, CA 94560, USA
e-mail: cgustafsson@dna20.com

S. Van Dien (ed.), *Metabolic Engineering for Bioprocess Commercialization*,
DOI 10.1007/978-3-319-41966-4_4

identified and their relative contribution quantified by iterative rounds of systematic exploration. The technology is generic and broadly applicable in biology and can be used within existing Quality by Design (QbD) processes to capture and interrogate design information far upstream of where QbD is typically applied for industrial scale bioprocesses. Several case studies that illustrate the efficiency and power of the approach are described.

4.1 Introduction

Microorganisms in nature produce a wide variety of diverse chemicals, though typically not at very high efficiencies. Metabolic engineering approaches help us direct and tune host metabolism to enable the production of large amounts of a molecule of interest. Early metabolic engineering endeavors were restricted due to lack of molecular biology and genetic engineering tools. Slow progress was made by cut and paste isolation assembly, and transplantation of natural DNA elements, genes and pathways into more favorable heterologous hosts and through random mutagenesis and screening for host optimization (Fig. 4.1). The invention of the polymerase chain reaction (PCR) provided a step change in the ability to alter and create novel DNA assemblies. In the past several decades, revolutionary advancements in the fields of molecular and microbiology, along with the emergence of sophisticated high-throughput laboratory automation and powerful computational capabilities, have shifted the paradigm for metabolic engineering.

Gene synthesis and DNA assembly technologies have obviated the need for preexisting physical DNA templates and enabled the de novo creation of synthetic sequences with novel properties (Fig. 4.1). It is now economically feasible to read as well as write entire genomes. Our ability to synthesize genes, pathways, and even designer genomes in a rapid and cost-effective manner has become an indispensable part of the metabolic pathway engineering cycle. These advances have led to several recent metabolic engineering successes. Microbial cells have been converted into factories for the production of pharmaceuticals and other valuable chemical products including the promising anticancer drug Taxol (Ajikumar et al. 2010), several terpenoids (Martin et al. 2003) including precursors to the anti-malarial drug Arteminisinin (Ro et al. 2006; Westfall et al. 2012; Paddon et al. 2013), 1-3 propanediol or Bio-PDO jointly by Dupont and Tate and Lyle (Nakamura and Whited 2003), 1-4 butanediol or BDO by Genomatica (Van Dien 2013; Yim et al. 2011), branched chain higher alcohols (Atsumi et al. 2008), alkanes (Schirmer et al. 2010; Choi and Lee 2013), as well as butyric acid and polylactic acid (Jung and Lee 2011; Jung et al. 2010; Jang et al. 2014).

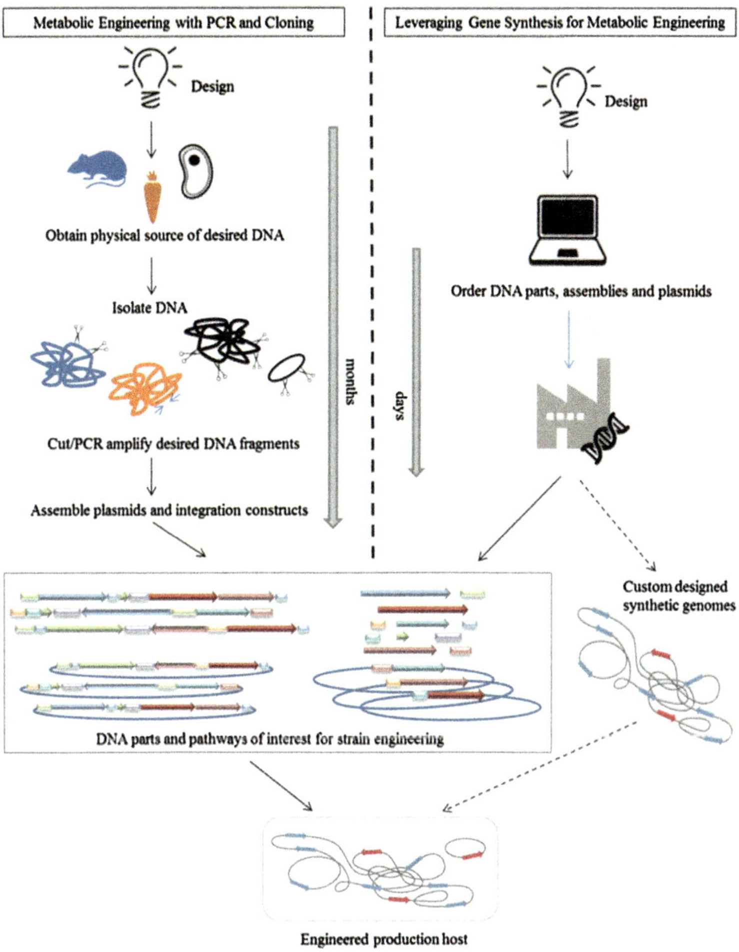

Fig. 4.1 Leveraging gene synthesis for present day metabolic engineering. Before the advent of gene synthesis technologies, scientists were confined to working with naturally occurring DNA sequences requiring availability of physical DNA templates (*left*). PCR-based tools facilitated minor changes (point mutations, promoter, and RBS variations) and enabled limited assembly of sequences from a variety of sources. The ability to synthesize any imaginable sequence enables the creation of DNA parts with novel functionalities and allows facile assembly of larger operons and pathways without requiring DNA templates (*right*). This now means that any previously characterized DNA sequence information that exists in electronic form (such as from a metagenomic effort) is now relatively simple to synthesize and evaluate in the context of a metabolic engineering endeavor. The bottleneck has shifted from synthesis to design

4.2 The Metabolic Pathway Engineering Cycle

4.2.1 Pathway Engineering Strategies

Metabolic engineering of pathways typically involves generating diversity through rational, semi-rational, and random approaches, followed by the evaluation of candidates for certain properties of interest. Rational methodologies focus on making precise, hypothesis-driven alterations to translate into improvements in production metrics; for example, altering the properties or expression levels of pathway enzymes. Random approaches rely on high throughput screens or clever selections to help identify candidates with improved characteristics. Semi-rational approaches combine elements of both approaches by leveraging knowledge about the system to reduce search space; e.g., hypothesis-driven decisions on which genes to target coupled with promoter or RBS libraries to perturb expression levels of each of these genes.

In addition to directly controlling pathway enzymes, there are whole-system approaches to influence pathway yields and fluxes. These include controlling concentrations of substrates and products, redox, cofactors, shunt products, and competing pathways. Direct and indirect influence on a particular metabolic pathway can be achieved by affecting the presence, abundance, and properties of certain proteins (Fig. 4.2). Genes can be inactivated, modified, or deleted. Heterologous genes can be introduced at permissive sites. The expression level of a gene can be modulated using a number of levers such as promoters, enhancers, activators, and ribosome-binding sites to ultimately impact the abundance of protein. The properties of an

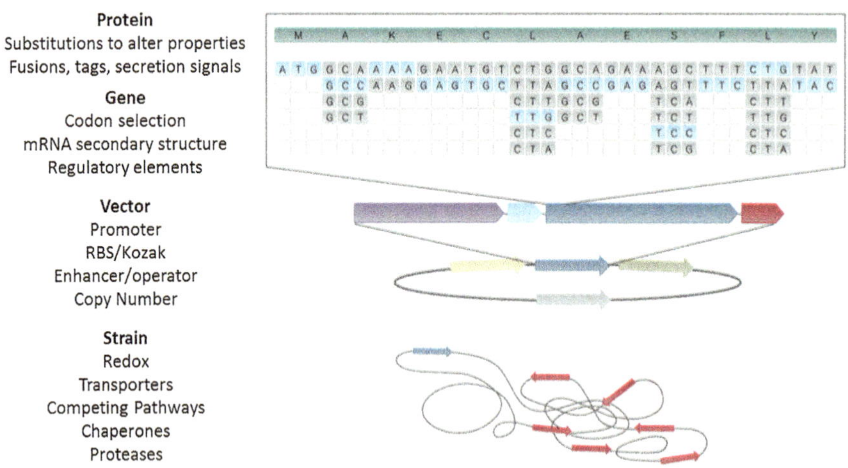

Fig. 4.2 Variable selection for metabolic pathway engineering. In addition to fermentation conditions which can be categorized as environmental variables, there are several other classes of variables that can all be controlled at the DNA level. Gene synthesis enables the effective exploration of this DNA sequence space in service of pathway optimization

enzyme in question can be altered using a multitude of protein engineering approaches. Conceptually, one can approach DNA sequence space at different levels of abstraction: codon, gene, promoter, pathway, transcriptional regulon, and whole genome. Fundamentally, these are all DNA base-pair level levers to influence a bio-system. Gene synthesis empowers researchers with the ability to de novo synthesize and modify biological components, pathways and genomes for metabolic engineering applications.

4.2.2 Gene Synthesis Technologies for Generating Diversity

Our knowledge and understanding of biological components such as enzymes and regulatory elements along with our ability to manipulate these components creates a means to derive a desired outcome in a given biological context (for example, overexpression of a protein or overproduction of a metabolite of interest). Several enabling technologies for the synthesis of genes, pathways, and whole genomes have emerged over the past 3 decades. Gene synthesis removes the requirement for naturally occurring DNA templates and provides the ability to synthesize any conceivable DNA sequence. This powerful technology is now integral to the field of metabolic engineering. In fact, synthetic DNA synthesis and assembly capabilities have led to the creation of the emerging field of synthetic biology. With the rapid time scales and manageable synthesis costs, we have the ability to create new DNA parts and assemblies with novel properties and functions, such as a variety of synthetic promoters and ribosome-binding sites (Mutalik et al. 2013a, b), transcriptional terminators (Chen et al. 2013), de novo designed enzymes (Jiang et al. 2008; Rothlisberger et al. 2008), biological logic circuits, and entire biosynthetic pathways (Menzella et al. 2005).

Since the first demonstration of gene synthesis of tRNA structural genes (Khorana et al. 1972), we have come a long way in extending and improving this capability. Several reports describing the synthesis of genes and plasmids followed (Itakura et al. 1977; Dillon and Rosen 1990; Mandecki et al. 1990; Stemmer et al. 1995). Smith et al. (2003) reported the assembly of the ~5.3 kb phiX174 bacteriophage genome from chemically synthesized oligonucleotides, where the purification of the desired final product required biological selection.

The early gene synthesis endeavors utilized ligation-based assembly approaches (Khorana 1979; Itakura et al. 1977). PCR-based polymerase cycling assembly or PCA (Smith et al. 2003), which involves the synthesis of oligonucleotides followed by ligation or PCR-based assembly of the final construct from an oligonucleotide pool, has proven to be very powerful in subsequent efforts. PCA is most reliable for the assembly of fragments of <1 kb in length. Larger fragments require more oligonucleotides to assemble into the final construct, and the resulting increased propensity for mis-priming tends to reduce assembly efficiencies. Furthermore, due to the common occurrence of errors during oligonucleotide synthesis (most predominantly single base or "n−1" deletions), large genes constructed by PCA from oligonucleotide pools most often also contained these errors. In cases where the genes being synthesized were essential for growth or propagation [e.g., synthesis of infectious phage DNA (Smith et al. 2003)], this biological selection enabled the identification of rare error-free clones. In the absence of a biological selection the identification of a clone with the intended sequence requires screening of a large number of candidates, a number which scales with the size of the fragment being synthesized. The use of error correction methodologies such as mismatch binding (Modrich 1991; Carr et al. 2004) and mismatch cleavage (Fuhrmann et al. 2005) following PCA helps alleviate the screening burden to some extent, though the issue of assembly efficiency for assembling a large number of oligonucleotides by PCA remains a challenge.

Most reports of the synthesis of large DNA constructs (>5 kb) describe the initial synthesis of smaller fragments or synthons by PCA, followed by assembly into the desired larger final construct. In 2002, Cello et al. described the construction of a 7.5 kb poliovirus cDNA (Cello et al. 2002). In 2004, Santi and colleagues at Kosan Biosciences described the assembly of a 32 kb polyketide synthase gene cluster for the production of erythromycin precursor 6-deoxy erythronolide B (Kodumal et al. 2004). With the total chemical synthesis of the approximately 600 kb genome of *M. genitalium* (Gibson et al. 2008a) and the 1 Mb genome of *M. mycoides* (Gibson et al. 2010), J. Craig Venter and coworkers created compelling technologies for the construction of very large DNA assemblies. These methods for constructing and altering DNA sequences at the genome scale (Lartigue et al. 2007, 2009; Benders et al. 2010) are very useful in the synthesis and manipulation of DNA constructs of all sizes, including sequences corresponding to large metabolic pathways.

As we develop the means to create larger DNA constructs in a rapid, robust, and cost-effective manner, several key enabling technologies have emerged. These include technologies for high quality, small volume DNA microchip-based oligonucleotide synthesis (Richmond et al. 2004; Tian et al. 2004; Zhou et al. 2004) and high fidelity polymerases such as the Phusion DNA polymerase from Thermo Fisher Scientific (Waltham, MA) with error rates lower than one in a million base pairs. This enables robust, accurate PCA of gene fragments from oligonucleotides, and when coupled with methods for error correction (Modrich 1991; Carr et al. 2004; Fuhrmann et al. 2005), enables facile generation of multi-kb constructs in a single step from oligonucleotides. In addition to these important technological advancements, efficient and convenient DNA assembly tools are critical for the construction of larger DNA fragments such as expression units, multigene pathways and even designer genomes.

Assembly of DNA fragments was first made possible by means of digestion by restriction enzymes and ligation of compatible ends. This method is severely limited in the number of fragments that can be assembled at a time and constrained to a high degree by the availability of unique restriction sites to generate compatible ends in the fragments being assembled. The discovery of type IIs restriction enzymes, which bind and cut double stranded DNA outside of their recognition sequences, led to the development of schemes including "Golden Gate Cloning" (Engler et al. 2008; Whitman et al. 2013) for the assembly of fragments at any desired junction. However, this approach is still limited in the number of fragments that can be assembled in a single step and is subject to the constraint that the sites used to linearize individual fragments cannot be present elsewhere in the sequences being assembled.

USER (uracil-specific *excision reagent*) cloning removes most sequence constraints and enables assembly of fragments with 6–10 bases of homology between fragments, but requires the PCR amplification of fragments with Uracil-containing amplification primers (Annaluru et al. 2012). Ligase cycling reaction (LCR) enables DNA assembly by means of bridging oligonucleotide sequences that are complementary to the ends of the DNA fragments being coupled along with a thermostable DNA ligase (Pachuk et al. 2000). Fragments for LCR assembly require 5′ phosphate groups and are joined end-to-end without requiring any overlaps between fragments.

Several methods that rely on ~30–50 bp overlaps between fragments have emerged. These methods typically require PCR for the generation of parts with overlapping regions for assembly. Circular polymerase extension cloning (CPEC) uses PCR to assemble multiple fragments in any vector (Quan and Tian 2009). Gibson isothermal assembly uses a 5′ exonuclease, a DNA polymerase and a DNA ligase for assembling multiple overlapping DNA fragments (Gibson et al. 2009). SLIC (Sequence and Ligation Independent Cloning) creates single-stranded overhangs in DNA fragments and vector sequences by means of an exonuclease and these fragments are then assembled in vitro (Li and Elledge 2012). SLiCE (Seamless Ligation Cloning Extract) enables DNA assembly in *E. coli* extracts made from cells expressing lambda phage recombinase (Zhang et al. 2012). Arguably the most powerful demonstrated method for assembly of large multi-gene pathways and chromosomes is in vivo assembly by homologous recombination in yeast (Shao et al. 2009; Gibson et al. 2008b).

Clearly there are several very useful methodologies for the assembly and manipulation of metabolic pathways, each with unique attributes (Cobb et al. 2014). The choice of method or combination of methods to implement will depend on the application, where factors such as throughput, modularity and flexibility in design, ease of use, and amenability to automation, number of fragments to be assembled, as well as properties of the sequences being assembled (length in base pairs, repetitive elements, GC content) need to be considered.

Modular design strategies can be coupled with the above assembly tools for combinatorial construction of multi-gene metabolic pathways, thereby maximizing diversity with a manageable number of parts. Several examples of clever modular design schemes have been reported. These include BioBricks (Shetty et al. 2008), 2ab assembly (Anderson et al. 2010), GoldenBraid (Sarrion-Perdigones et al. 2011), MoClo (Werner et al. 2012; Weber et al. 2011), MODAL (Casini et al. 2014), as well as the rapid yeast strain engineering (RYSE) linkers developed at Amyris (Serber et al. 2012; de Kok et al. 2014).

While gene synthesis and DNA assembly methods enable the convenient construction of genes, pathways, and entire chromosomes, it is often desirable to alter specific genes or genomic regions in a host chromosome (deletions, point mutations, replacements). A tremendous amount of progress has been made with classical genetic engineering tools that rely solely on homologous recombination. More recent advances in genome engineering technologies have been shown to work in a wide variety of hosts with great efficiency and speed (Esvelt and Wang 2013). The ability to make precise genome-wide modifications has relieved a critical bottleneck in the metabolic pathway engineering cycle.

4.2.3 Assay and Analysis

For a given metabolic pathway engineering endeavor, there are a multitude of biological components that need to be explored in order to navigate toward optimal performance. Our significant understanding of biological systems enables us to readily identify a number of variables that are likely to impact the performance of the pathway in question. There are several examples reporting the use of rational design and low to medium throughput assay methodologies to achieve tremendous improvements in the production of a molecule of interest in a variety of hosts.

Gene synthesis technologies have greatly enhanced our abilities to engineer biological systems. First, it is now possible to synthesize any desired sequence, enabling the generation of DNA parts with novel functions, and adaptation of natural components for a heterologous host (e.g., codon optimization of genes). Gene synthesis enables the refactoring and modularization of metabolic pathways to enable convenient evaluation of the impact of individual components in the pathway. In addition, the access to fast and efficient DNA synthesis at the parts, pathway, and genome scales without requiring physical DNA templates accelerates the process of diversity generation in the metabolic engineering cycle. These advantages are illustrated in the gene synthesis and redesign of naturally occurring polyketide synthase genes and heterologous production of a variety of novel polyketides in *E. coli* (Kodumal et al. 2004; Menzella et al. 2005). These themes of adaptation of genes for heterologous expression as well as modularization of biosynthesis pathways for easy manipulation of individual components are now fairly common practice (Paddon et al. 2013; Westfall et al. 2012).

Despite the examples of progress made with rational and semi-rational approaches, it would be remiss not to acknowledge that there is still a tremendous amount we do not yet understand. This is why most if not all commercial efforts for strain improvement include a random mutagenesis and screening component. Mutagenesis and screening efforts require higher throughput since the possible diversity space being explored is now significantly expanded. Further, since no library is ever perfectly random, oversampling is needed to maximize the number of variables being sampled (Levay-Young et al. 2013). Our ability to screen has been greatly enhanced by automation and high throughput capabilities that enable the miniaturization of assays to microplates and nanoliter scale droplets. While random mutagenesis and screening efforts have the potential to identify unanticipated beneficial mutations with very low direct costs for generating individual mutants, they require significant effort and screening capabilities, and often lead to no insights on how to improve the system further.

At DNA2.0 we believe that the most valuable aspect of the engineering cycle is the knowledge gained from a screening effort. Learning about the impact of individual variables on the output property of interest can enable rapid navigation toward an optimal solution. Metabolic pathway engineering is a multivariate problem involving a multitude of variables, at the gene, protein, pathway, and host genome level (Fig. 4.2). In addition, there are environmental variables that one can manipulate to achieve a desired outcome. Attempting to explore different levels for each of these variables, as well as all of the combinations thereof, is prohibitive. To deal with this extremely large diversity space, we employ an approach that combines (i) efficient exploration of variables through Design of Experiments (Fisher 1935) (ii) precise construction of variant strains for sampling by means of gene synthesis and genome modification, along with (iii) multivariate data analysis (MVDA) (Hand et al. 2001) tools to navigate towards an optimal solution (Fig. 4.3).

A Design of Experiments method, which works by maximizing the number of variables evaluated in every test while minimizing co-variation between these variables, can be used to efficiently navigate this landscape. The approach was pioneered in 1993 with the design of an *E. coli* promoter (Jonsson et al. 1993) but was at the time limited by poor access to gene synthesis. Today gene synthesis and modern genome editing tools enable the precise construction of every necessary gene, pathway, and strain genotype with the necessary combinations of modification for evaluation. MVDA methods such as Neural Net (NN), Support Vector Machine (SVM), and Partial Least Squares (PLS) regression modeling enable us to identify causal variables, from which beneficial changes can be combined in an iterative process to navigate towards a system optimum.

This approach enables the exploration a large number of variables, each in multiple contexts with a relatively small number of samples to evaluate. This means that one can assess a large number of variables with only two to three times that number in measurements to be performed. The small number of samples required offers unique advantages. First, this can eliminate the need for a surrogate assay since it is now possible to perform the highest quality and most relevant end-point assay instead. High quality assays mean high precision from improved signal and reduced noise, which in turn improve the quality of the data and therefore the accuracy of the resulting predictive model. In addition, one can make multiple measurements on the same samples, thereby enabling the simultaneous optimization for multiple desirable characteristics. For example under different production conditions, one might collect information on pathway products and intermediates (productivity, yields, titers, fluxes) in addition to production host fitness and growth rate data.

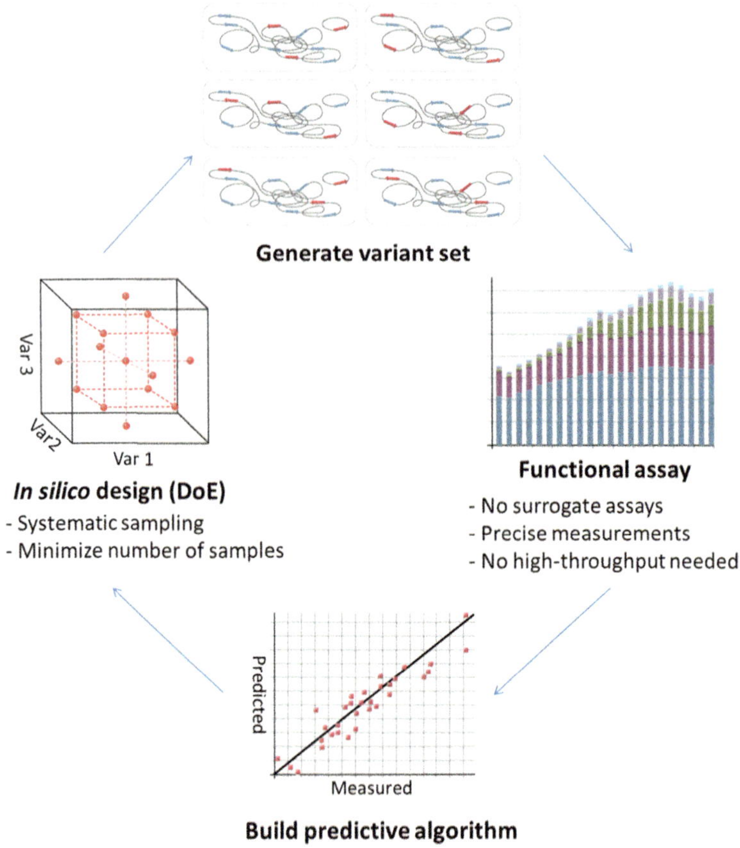

Fig. 4.3 A whole-system optimization approach to metabolic pathway engineering. Whether optimizing genes for better protein expression, engineering proteins with desirable properties, or creating designer pathways and production strains for generating a wide variety of products ranging from fine chemicals to protein pharmaceuticals, the process is iterative and each round can be broken down into the following steps; (i) Determination of variables to explore and DOE to design variant set, (ii) Construction of variants, (iii) Evaluation of variants for properties of interest, (iv) Construction of predictive model from which we can determine beneficial variables. Beneficial variables are enriched in subsequent round(s), usually leading to candidates that meet or exceed desired production metrics

A predictive model enables the identification of patterns in data that can be used to predict the effects of multiple input variables on a particular output variable. In the case of metabolic pathways, pathway optimization can be approached at the individual gene level, pathway level, or whole genome level. Figure 4.2 lists examples of input variables that can be explored to influence pathway performance.

Since models are approximations, it is important to assess their reliability. A model should be evaluated for its ability to fit the data used to construct it, as well as its ability to predict effects of salient input variables on the desired output variables. The former is assessed through a scoring function. The scoring function is typically determined after performing an experiment where an output of interest is measured for different combinations of the variables that describe the system. The predictive power of a model can be assessed by cross-validation (CV). Here, experimentally generated data is split into two groups: a training set and a testing set. The model is constructed using the training set, which in turn is used to predict the testing set. A good predictive model will accurately predict the testing set. Properly constructed models are able to consider multiple factors and predict outcomes with a high level of accuracy. Figure 4.4 is an example of a model constructed for various amino acid substitutions in the sequence of a Glutathione S-transferase (GST) enzyme, in an effort to improve its activity on a particular herbicide substrate (Metolachlor). A good agreement between the predicted and measured values for the influence of every variable with R^2 and R^2 (CV) values approaching a value of 1 is indicative of a consistent and strong predictive model.

An important output of a predictive regression model is a set of regression weights assigned to the individual variables. These regression weights are a measure of the impact of a particular variable on the system. Variables that are predicted

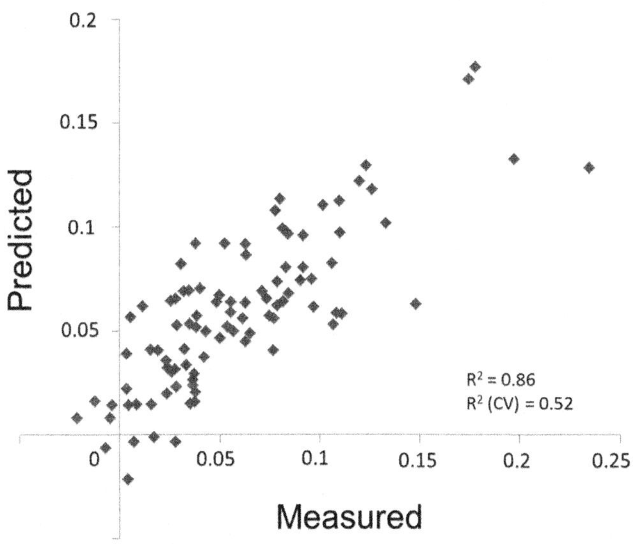

Fig. 4.4 An example of a model constructed for altering the substrate specificity of a Glutathione S-transferase (GST) enzyme. Various amino acid substitutions in the GST sequence were evaluated in an effort to improve its activity on a particular substrate (Metolachlor). A variant set (where each variant sequence contains a unique combination of multiple amino acid substitutions) is constructed and evaluated for its activity on the substrate of interest. The predictive model is constructed on the basis of this measured data. The R^2 value is an indicator of how well the data fits the model. Next, for cross-validation of the model, experimentally generated data is split into two groups: a training set and a testing set. The model is constructed using the training set, which in turn is used to predict the testing set. A good predictive model will accurately predict the testing set, which is indicated by an R^2 (CV) value that approaches 1

to be beneficial based on regression weights are enriched in subsequent rounds of optimization.

We routinely leverage our gene synthesis and machine learning capabilities at DNA2.0 to optimize genes for better protein expression, to engineer proteins with desirable properties, as well as to create designer pathways and production strains for the production of a variety of products ranging from fine chemicals to pharmaceutical proteins. Regardless of the application, the optimization process is iterative and the individual steps are essentially the same in principle (Fig. 4.3).

One critical aspect of an optimization endeavor is variable selection. In the context of optimizing the DNA sequence of a gene for protein expression, common variables that we explore include codon usage frequencies for each amino acid, RNA secondary structure parameters (e.g., predicted free energy of folding in the translation initiation region), GC content, and translation initiation (Welch et al. 2009b; Gustafsson et al. 2012). Gene synthesis also allows us to conveniently avoid regulatory motifs and other undesirable sequences including particular restriction sites. Figure 4.5a shows regression weights calculated for different codon usage frequencies for a set of dasher GFP protein variants in 2 different production hosts. In each case, we were able to develop a strong predictive model which showed that synonymous codon substitution had a large effect on protein production. We are able to utilize similar models for a variety of different production hosts to design and construct genes for over-expression of any protein of interest in that particular host system (Welch et al. 2009a; Mellitzer et al. 2014).

When optimizing proteins for a particular property of interest, variables consist of amino acid substitutions at various positions in that protein. We identify potential substitutions for a given protein by first querying its sequence against Genbank using BLAST. The sequences of identified homologues are aligned using ClustalW and all naturally occurring amino acids at every position in the reference sequence are recorded. Substitutions are chosen from this list based on a variety of different criterion as described previously (Liao et al. 2007; Govindarajan et al. 2014). Gene synthesis enables the precise and rapid synthesis of every variant gene for evaluation. Variant proteins are assayed for properties of interest and a model is created to identify beneficial substitutions and combinations thereof. The knowledge gained from one round is applied to create variants with desired properties in subsequent rounds. Using this approach, one can simultaneously optimize for multiple properties of interest. Figure 4.5b shows regression weights calculated for various amino acid substitutions in a GST enzyme, as a measure of their impact on enzyme activity towards two novel herbicide substrates. This methodology has led to scores of successful examples of model-guided improvement of proteins in an industrially relevant context. Examples include the 60-fold improvement of the initial velocity of a wild-type Vfat enzyme (Midelfort et al. 2013), development of glutathione transferases with activity towards a diverse set of herbicides (Govindarajan et al. 2014), engineering of a thermostable Proteinase K (Liao et al. 2007), thermostable cellulases (Heinzelman et al. 2009), and the identification of prolyl endopeptidase variants with enhanced specific activities at low pH as well as 200-fold resistance to pepsin (Ehren et al. 2008). As illustrated in Fig. 4.6, we were able to utilize this

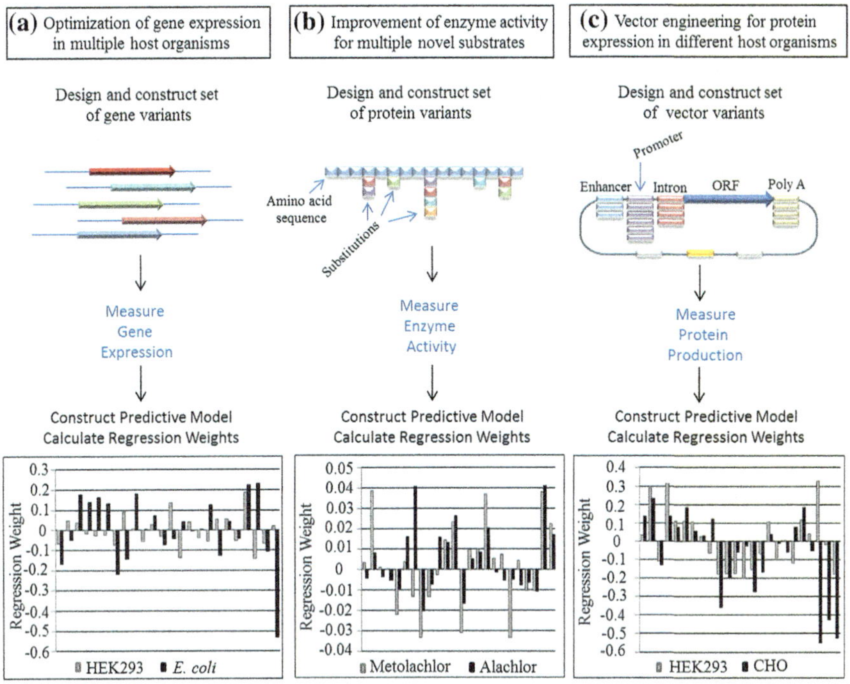

Fig. 4.5 Determination of regression weights as a measure of the impact of individual variables on output variables of interest. Below are regression weights calculated for **a** the impact of different codon usage frequencies on dasher GFP protein expression in two different hosts (*E. coli = black bars*, HEK293 cells = *gray bars*), **b** the impact of various amino acid substitutions on the activity of a GST enzyme on two different substrates (Alachlor = *black bars*, Metolachlor = *gray bars*), **c** the impact of various expression elements such as promoters and enhancers on dasher GFP expression in two different mammalian expression hosts (CHO cells = *black bars*, HEK293 cells = *gray bars*). In each case, the regression weights for 24 individual variables are shown. These examples illustrate that we can construct models for experiments conducted in different contexts using the same set of variants. We find that the impact of a particular input variable can vary significantly based on the assay conditions as well as the output variable of interest

approach to improve the activity of an enzyme by 14-fold with only 192 variants screened over 3 rounds.

We have also successfully used this approach to develop a set of vectors for the optimization of protein expression. Vectors were constructed featuring combinations of sequence elements including enhancers, promoters, introns, polyA motifs, and amplification elements. The vectors were used to express DasherGFP in HEK293 cells and CHO cells. Because the vector components were varied systematically, we were able to use machine learning to assess the contribution of each element to protein expression in HEK293 as well as CHO cells (Fig. 4.5c). Most promoters and polyA sequences performed similarly in the two systems.

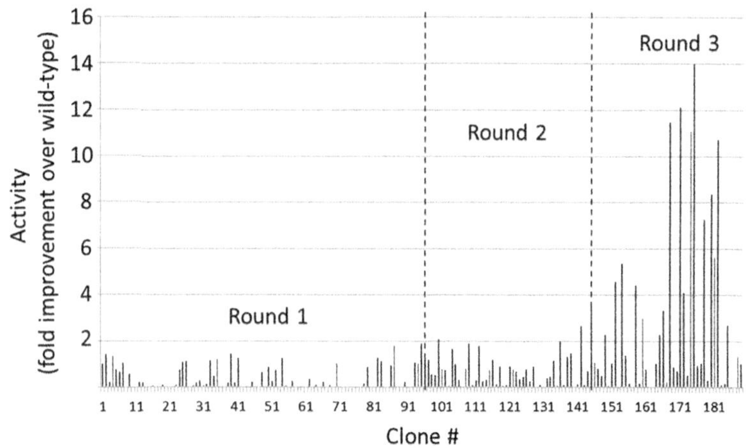

Fig. 4.6 Rapid progress and efficient exploration. This iterative optimization approach was utilized to improve the activity of a model enzyme 14-fold over wild-type with only 192 variants screened over 3 rounds

Interestingly, the viral amplification sequences tested were found to have a variable effect in HEK293 cells and strongly negative effect in CHO cells.

Similarly, this approach can be extended to the optimization of pathways and strains for the production of a metabolite of interest or for the production of a pharmaceutically relevant protein therapeutic. Gene synthesis enables the creation of precise variants for evaluation. DOE and machine learning provide efficient and effective means for exploration of the vast and multidimensional biological diversity space.

4.3 Discussion

The ability to synthesize and alter genes, pathways and even genomes offers unique capabilities to any bioengineering endeavor. Variables, from codon choice to amino acid sequence of a protein to promoters driving a set of genes in a genome, can all be controlled at the DNA level. Gene synthesis empowers metabolic engineers with the ability to create DNA parts with novel properties, adapt components to function in heterologous hosts, and to refactor biosynthesis pathways for convenient manipulation.

We believe that emphasis should be placed on the *learn* aspect of the metabolic pathway engineering (design-build-test-learn) cycle and make the case for efficient systematic exploration of the whole genome space to identify beneficial variables. Gene synthesis technologies enable the precise synthesis of exactly the desired information rich sequences that will most efficiently sample the diversity space and guide the advancement towards an optimal solution. The small number of samples enables the evaluation of variants using the most relevant endpoint assay without requiring compromises in quality or relevance on account of cost or throughput. This approach also enables simultaneous optimization of multiple attributes of interest. Multiple examples of the valuable information and demonstrable progress that is generated from this approach in a number of different contexts illustrate the power and wide applicability of this approach. Regardless of the rationale for metabolic pathway optimization, gene synthesis forms an essential part of today's metabolic pathway engineering process.

References

Ajikumar PK, Xiao WH, Tyo KE, Wang Y, Simeon F, Leonard E, Mucha O, Phon TH, Pfeifer B, Stephanopoulos G (2010) Isoprenoid pathway optimization for Taxol precursor overproduction in Escherichia coli. Science 330(6000):70–74. doi:10.1126/science.1191652

Anderson JC, Dueber JE, Leguia M, Wu GC, Goler JA, Arkin AP, Keasling JD (2010) BglBricks: a flexible standard for biological part assembly. J Biol Eng 4(1):1. doi:10.1186/1754-1611-4-1

Annaluru N, Muller H, Ramalingam S, Kandavelou K, London V, Richardson SM, Dymond JS, Cooper EM, Bader JS, Boeke JD, Chandrasegaran S (2012) Assembling DNA fragments by USER fusion. Methods Mol Biol 852:77–95. doi:10.1007/978-1-61779-564-0_7

Atsumi S, Hanai T, Liao JC (2008) Non-fermentative pathways for synthesis of branched-chain higher alcohols as biofuels. Nature 451(7174):86–89. doi:10.1038/nature06450

Benders GA, Noskov VN, Denisova EA, Lartigue C, Gibson DG, Assad-Garcia N, Chuang RY, Carrera W, Moodie M, Algire MA, Phan Q, Alperovich N, Vashee S, Merryman C, Venter JC, Smith HO, Glass JI, Hutchison CA 3rd (2010) Cloning whole bacterial genomes in yeast. Nucleic Acids Res 38(8):2558–2569. doi:10.1093/nar/gkq119

Carr PA, Park JS, Lee YJ, Yu T, Zhang S, Jacobson JM (2004) Protein-mediated error correction for de novo DNA synthesis. Nucleic Acids Res 32(20):e162. doi:10.1093/nar/gnh160

Casini A, MacDonald JT, De Jonghe J, Christodoulou G, Freemont PS, Baldwin GS, Ellis T (2014) One-pot DNA construction for synthetic biology: the Modular Overlap-Directed Assembly with Linkers (MODAL) strategy. Nucleic Acids Res 42(1):e7. doi:10.1093/nar/gkt915

Cello J, Paul AV, Wimmer E (2002) Chemical synthesis of poliovirus cDNA: generation of infectious virus in the absence of natural template. Science 297(5583):1016–1018. doi:10.1126/science.1072266

Chen YJ, Liu P, Nielsen AA, Brophy JA, Clancy K, Peterson T, Voigt CA (2013) Characterization of 582 natural and synthetic terminators and quantification of their design constraints. Nat Methods 10(7):659–664. doi:10.1038/nmeth.2515

Choi YJ, Lee SY (2013) Microbial production of short-chain alkanes. Nature 502(7472):571–574. doi:10.1038/nature12536

Cobb RE, Ning JC, Zhao H (2014) DNA assembly techniques for next-generation combinatorial biosynthesis of natural products. J Ind Microbiol Biotechnol 41(2):469–477. doi:10.1007/s10295-013-1358-3

de Kok S, Stanton LH, Slaby T, Durot M, Holmes VF, Patel KG, Platt D, Shapland EB, Serber Z, Dean J, Newman JD, Chandran SS (2014) Rapid and reliable DNA assembly via ligase cycling reaction. ACS Synth Biol 3(2):97–106. doi:10.1021/sb4001992

Dillon PJ, Rosen CA (1990) A rapid method for the construction of synthetic genes using the polymerase chain reaction. Biotechniques 9(3):298, 300

Ehren J, Govindarajan S, Moron B, Minshull J, Khosla C (2008) Protein engineering of improved prolyl endopeptidases for celiac sprue therapy. Protein Eng Des Sel: PEDS 21(12):699–707. doi:10.1093/protein/gzn050

Engler C, Kandzia R, Marillonnet S (2008) A one pot, one step, precision cloning method with high throughput capability. PLoS ONE 3(11):e3647. doi:10.1371/journal.pone.0003647

Esvelt KM, Wang HH (2013) Genome-scale engineering for systems and synthetic biology. Mol Syst Biol 9:641. doi:10.1038/msb.2012.66

Fisher RA (1935) Design of experiments. Oliver and Boyd, London

Fuhrmann M, Oertel W, Berthold P, Hegemann P (2005) Removal of mismatched bases from synthetic genes by enzymatic mismatch cleavage. Nucleic Acids Res 33(6):e58. doi:10.1093/nar/gni058

Gibson DG, Benders GA, Andrews-Pfannkoch C, Denisova EA, Baden-Tillson H, Zaveri J, Stockwell TB, Brownley A, Thomas DW, Algire MA, Merryman C, Young L, Noskov VN, Glass JI, Venter JC, Hutchison CA 3rd, Smith HO (2008a) Complete chemical synthesis, assembly, and cloning of a Mycoplasma genitalium genome. Science 319(5867):1215–1220. doi:10.1126/science.1151721

Gibson DG, Benders GA, Axelrod KC, Zaveri J, Algire MA, Moodie M, Montague MG, Venter JC, Smith HO, Hutchison CA 3rd (2008b) One-step assembly in yeast of 25 overlapping DNA fragments to form a complete synthetic Mycoplasma genitalium genome. Proc Natl Acad Sci USA 105(51):20404–20409. doi:10.1073/pnas.0811011106

Gibson DG, Glass JI, Lartigue C, Noskov VN, Chuang RY, Algire MA, Benders GA, Montague MG, Ma L, Moodie MM, Merryman C, Vashee S, Krishnakumar R, Assad-Garcia N, Andrews-Pfannkoch C, Denisova EA, Young L, Qi ZQ, Segall-Shapiro TH, Calvey CH, Parmar PP, Hutchison CA 3rd, Smith HO, Venter JC (2010) Creation of a bacterial cell controlled by a chemically synthesized genome. Science 329(5987):52–56. doi:10.1126/science.1190719

Gibson DG, Young L, Chuang RY, Venter JC, Hutchison CA 3rd, Smith HO (2009) Enzymatic assembly of DNA molecules up to several hundred kilobases. Nat Methods 6(5):343–345. doi:10.1038/nmeth.1318

Govindarajan S, Mannervik B, Silverman JA, Wright K, Regitsky D, Hegazy U, Purcell TJ, Welch M, Minshull J, Gustafsson C (2014) Mapping of amino acid substitutions conferring herbicide resistance in wheat glutathione transferase. ACS Synth Biol. doi:10.1021/sb500242x

Gustafsson C, Minshull J, Govindarajan S, Ness J, Villalobos A, Welch M (2012) Engineering genes for predictable protein expression. Protein Expr Purif 83(1):37–46. doi:10.1016/j.pep.2012.02.013

Hand D, Mannila H, Smyth P (2001) Principles of data mining. The MIT Press, Cambridge, MA

Heinzelman P, Snow CD, Wu I, Nguyen C, Villalobos A, Govindarajan S, Minshull J, Arnold FH (2009) A family of thermostable fungal cellulases created by structure-guided recombination. Proc Natl Acad Sci USA 106(14):5610–5615. doi:10.1073/pnas.0901417106

Itakura K, Hirose T, Crea R, Riggs AD, Heyneker HL, Bolivar F, Boyer HW (1977) Expression in Escherichia coli of a chemically synthesized gene for the hormone somatostatin. Science 198(4321):1056–1063

Jang YS, Im JA, Choi SY, Lee JI, Lee SY (2014) Metabolic engineering of Clostridium acetobutylicum for butyric acid production with high butyric acid selectivity. Metab Eng 23:165–174. doi:10.1016/j.ymben.2014.03.004

Jiang L, Althoff EA, Clemente FR, Doyle L, Rothlisberger D, Zanghellini A, Gallaher JL, Betker JL, Tanaka F, Barbas CF 3rd, Hilvert D, Houk KN, Stoddard BL, Baker D (2008) De novo computational design of retro-aldol enzymes. Science 319(5868):1387–1391. doi:10.1126/science.1152692

Jonsson J, Norberg T, Carlsson L, Gustafsson C, Wold S (1993) Quantitative sequence-activity models (QSAM)–tools for sequence design. Nucleic Acids Res 21(3):733–739

Jung YK, Lee SY (2011) Efficient production of polylactic acid and its copolymers by metabolically engineered Escherichia coli. J Biotechnol 151(1):94–101. doi:10.1016/j.jbiotec.2010.11.009

Jung YK, Kim TY, Park SJ, Lee SY (2010) Metabolic engineering of Escherichia coli for the production of polylactic acid and its copolymers. Biotechnol Bioeng 105(1):161–171. doi:10.1002/bit.22548

Khorana HG (1979) Total synthesis of a gene. Science 203(4381):614–625

Khorana HG, Agarwal KL, Buchi H, Caruthers MH, Gupta NK, Kleppe K, Kumar A, Otsuka E, RajBhandary UL, Van de Sande JH, Sgaramella V, Terao T, Weber H, Yamada T (1972) Studies on polynucleotides. 103. Total synthesis of the structural gene for an alanine transfer ribonucleic acid from yeast. J Mol Biol 72(2):209–217

Kodumal SJ, Patel KG, Reid R, Menzella HG, Welch M, Santi DV (2004) Total synthesis of long DNA sequences: synthesis of a contiguous 32-kb polyketide synthase gene cluster. Proc Natl Acad Sci USA 101(44):15573–15578. doi:10.1073/pnas.0406911101

Lartigue C, Glass JI, Alperovich N, Pieper R, Parmar PP, Hutchison CA 3rd, Smith HO, Venter JC (2007) Genome transplantation in bacteria: changing one species to another. Science 317 (5838):632–638. doi:10.1126/science.1144622

Lartigue C, Vashee S, Algire MA, Chuang RY, Benders GA, Ma L, Noskov VN, Denisova EA, Gibson DG, Assad-Garcia N, Alperovich N, Thomas DW, Merryman C, Hutchison CA 3rd, Smith HO, Venter JC, Glass JI (2009) Creating bacterial strains from genomes that have been cloned and engineered in yeast. Science 325(5948):1693–1696. doi:10.1126/science.1173759

Levay-Young B, Olesiuk M, Gustafsson C, Minshull J (2013) Library format for bioengineering; maximizing screening efficiency through good design. In: Genetic engineering & biotechnology news, vol 33. Mary Ann Liebert, Inc., New Rochelle, NY

Li MZ, Elledge SJ (2012) SLIC: a method for sequence- and ligation-independent cloning. Methods Mol Biol 852:51–59. doi:10.1007/978-1-61779-564-0_5

Liao J, Warmuth MK, Govindarajan S, Ness JE, Wang RP, Gustafsson C, Minshull J (2007) Engineering proteinase K using machine learning and synthetic genes. BMC Biotechnol 7:16. doi:10.1186/1472-6750-7-16

Mandecki W, Hayden MA, Shallcross MA, Stotland E (1990) A totally synthetic plasmid for general cloning, gene expression and mutagenesis in Escherichia coli. Gene 94(1):103–107

Martin VJ, Pitera DJ, Withers ST, Newman JD, Keasling JD (2003) Engineering a mevalonate pathway in Escherichia coli for production of terpenoids. Nat Biotechnol 21(7):796–802. doi:10.1038/nbt833

Mellitzer A, Ruth C, Gustafsson C, Welch M, Birner-Grunberger R, Weis R, Purkarthofer T, Glieder A (2014) Synergistic modular promoter and gene optimization to push cellulase secretion by Pichia pastoris beyond existing benchmarks. J Biotechnol 191:187–195. doi:10.1016/j.jbiotec.2014.08.035

Menzella HG, Reid R, Carney JR, Chandran SS, Reisinger SJ, Patel KG, Hopwood DA, Santi DV (2005) Combinatorial polyketide biosynthesis by de novo design and rearrangement of modular polyketide synthase genes. Nat Biotechnol 23(9):1171–1176. doi:10.1038/nbt1128

Midelfort KS, Kumar R, Han S, Karmilowicz MJ, McConnell K, Gehlhaar DK, Mistry A, Chang JS, Anderson M, Villalobos A, Minshull J, Govindarajan S, Wong JW (2013) Redesigning and characterizing the substrate specificity and activity of Vibrio fluvialis aminotransferase for the synthesis of imagabalin. Protein Eng Des Sel: PEDS 26(1):25–33. doi:10.1093/protein/gzs065

Modrich P (1991) Mechanisms and biological effects of mismatch repair. Annu Rev Genet 25:229–253. doi:10.1146/annurev.ge.25.120191.001305

Mutalik VK, Guimaraes JC, Cambray G, Lam C, Christoffersen MJ, Mai QA, Tran AB, Paull M, Keasling JD, Arkin AP, Endy D (2013a) Precise and reliable gene expression via standard transcription and translation initiation elements. Nat Methods 10(4):354–360. doi:10.1038/nmeth.2404

Mutalik VK, Guimaraes JC, Cambray G, Mai QA, Christoffersen MJ, Martin L, Yu A, Lam C, Rodriguez C, Bennett G, Keasling JD, Endy D, Arkin AP (2013b) Quantitative estimation of activity and quality for collections of functional genetic elements. Nat Methods 10(4):347–353. doi:10.1038/nmeth.2403

Nakamura CE, Whited GM (2003) Metabolic engineering for the microbial production of 1,3-propanediol. Curr Opin Biotechnol 14(5):454–459. doi:10.1016/j.copbio.2003.08.005

Pachuk CJ, Samuel M, Zurawski JA, Snyder L, Phillips P, Satishchandran C (2000) Chain reaction cloning: a one-step method for directional ligation of multiple DNA fragments. Gene 243 (1–2):19–25

Paddon CJ, Westfall PJ, Pitera DJ, Benjamin K, Fisher K, McPhee D, Leavell MD, Tai A, Main A, Eng D, Polichuk DR, Teoh KH, Reed DW, Treynor T, Lenihan J, Fleck M, Bajad S, Dang G, Dengrove D, Diola D, Dorin G, Ellens KW, Fickes S, Galazzo J, Gaucher SP, Geistlinger T, Henry R, Hepp M, Horning T, Iqbal T, Jiang H, Kizer L, Lieu B, Melis D, Moss N, Regentin R, Secrest S, Tsuruta H, Vazquez R, Westblade LF, Xu L, Yu M, Zhang Y, Zhao L, Lievense J, Covello PS, Keasling JD, Reiling KK, Renninger NS, Newman JD (2013) High-level semi-synthetic production of the potent antimalarial artemisinin. Nature 496 (7446):528–532. doi:10.1038/nature12051

Quan J, Tian J (2009) Circular polymerase extension cloning of complex gene libraries and pathways. PLoS ONE 4(7):e6441. doi:10.1371/journal.pone.0006441

Richmond KE, Li MH, Rodesch MJ, Patel M, Lowe AM, Kim C, Chu LL, Venkataramaian N, Flickinger SF, Kaysen J, Belshaw PJ, Sussman MR, Cerrina F (2004) Amplification and assembly of chip-eluted DNA (AACED): a method for high-throughput gene synthesis. Nucleic Acids Res 32(17):5011–5018. doi:10.1093/nar/gkh793

Ro DK, Paradise EM, Ouellet M, Fisher KJ, Newman KL, Ndungu JM, Ho KA, Eachus RA, Ham TS, Kirby J, Chang MC, Withers ST, Shiba Y, Sarpong R, Keasling JD (2006) Production of the antimalarial drug precursor artemisinic acid in engineered yeast. Nature 440 (7086):940–943. doi:10.1038/nature04640

Rothlisberger D, Khersonsky O, Wollacott AM, Jiang L, DeChancie J, Betker J, Gallaher JL, Althoff EA, Zanghellini A, Dym O, Albeck S, Houk KN, Tawfik DS, Baker D (2008) Kemp elimination catalysts by computational enzyme design. Nature 453(7192):190–195. doi:10.1038/nature06879

Sarrion-Perdigones A, Falconi EE, Zandalinas SI, Juarez P, Fernandez-del-Carmen A, Granell A, Orzaez D (2011) GoldenBraid: an iterative cloning system for standardized assembly of reusable genetic modules. PLoS ONE 6(7):e21622. doi:10.1371/journal.pone.0021622

Schirmer A, Rude MA, Li X, Popova E, del Cardayre SB (2010) Microbial biosynthesis of alkanes. Science 329(5991):559–562. doi:10.1126/science.1187936

Serber Z, Lowe R, Ubersax J, Chandran S (2012) Compositions and methods for the assembly of polynucleotides. 8,110,360 B2, Feb 7, 2012

Shao Z, Zhao H, Zhao H (2009) DNA assembler, an in vivo genetic method for rapid construction of biochemical pathways. Nucleic Acids Res 37(2):e16. doi:10.1093/nar/gkn991

Shetty RP, Endy D, Knight TF Jr (2008) Engineering BioBrick vectors from BioBrick parts. J Biol Eng 2:5. doi:10.1186/1754-1611-2-5

Smith HO, Hutchison CA 3rd, Pfannkoch C, Venter JC (2003) Generating a synthetic genome by whole genome assembly: phiX174 bacteriophage from synthetic oligonucleotides. Proc Natl Acad Sci USA 100(26):15440–15445. doi:10.1073/pnas.2237126100

Stemmer WP, Crameri A, Ha KD, Brennan TM, Heyneker HL (1995) Single-step assembly of a gene and entire plasmid from large numbers of oligodeoxyribonucleotides. Gene 164(1):49–53

Tian J, Gong H, Sheng N, Zhou X, Gulari E, Gao X, Church G (2004) Accurate multiplex gene synthesis from programmable DNA microchips. Nature 432(7020):1050–1054. doi:10.1038/nature03151

Van Dien S (2013) From the first drop to the first truckload: commercialization of microbial processes for renewable chemicals. Curr Opin Biotechnol 24(6):1061–1068. doi:10.1016/j.copbio.2013.03.002

Weber E, Engler C, Gruetzner R, Werner S, Marillonnet S (2011) A modular cloning system for standardized assembly of multigene constructs. PLoS ONE 6(2):e16765. doi:10.1371/journal.pone.0016765

Welch M, Govindarajan S, Ness JE, Villalobos A, Gurney A, Minshull J, Gustafsson C (2009a) Design parameters to control synthetic gene expression in Escherichia coli. PLoS ONE 4(9): e7002. doi:10.1371/journal.pone.0007002

Welch M, Villalobos A, Gustafsson C, Minshull J (2009b) You're one in a googol: optimizing genes for protein expression. J R Soc Interface/R Soc 6(Suppl 4):S467–S476. doi:10.1098/rsif.2008.0520.focus

Werner S, Engler C, Weber E, Gruetzner R, Marillonnet S (2012) Fast track assembly of multigene constructs using Golden Gate cloning and the MoClo system. Bioeng Bugs 3(1):38–43. doi:10.4161/bbug.3.1.18223

Westfall PJ, Pitera DJ, Lenihan JR, Eng D, Woolard FX, Regentin R, Horning T, Tsuruta H, Melis DJ, Owens A, Fickes S, Diola D, Benjamin KR, Keasling JD, Leavell MD, McPhee DJ, Renninger NS, Newman JD, Paddon CJ (2012) Production of amorphadiene in yeast, and its conversion to dihydroartemisinic acid, precursor to the antimalarial agent artemisinin. Proc Natl Acad Sci USA 109(3):E111–E118. doi:10.1073/pnas.1110740109

Whitman L, Gore M, Ness J, Theodorou E, Gustafsson C, Minshull J (2013) Rapid, scarless cloning of gene fragments using the electra vector system™. In: Genetic engineering & biotechnology news, vol 33. Mary Ann Liebert, Inc., New Rochelle, NY

Yim H, Haselbeck R, Niu W, Pujol-Baxley C, Burgard A, Boldt J, Khandurina J, Trawick JD, Osterhout RE, Stephen R, Estadilla J, Teisan S, Schreyer HB, Andrae S, Yang TH, Lee SY, Burk MJ, Van Dien S (2011) Metabolic engineering of Escherichia coli for direct production of 1,4-butanediol. Nat Chem Biol 7(7):445–452. doi:10.1038/nchembio.580

Zhang Y, Werling U, Edelmann W (2012) SLiCE: a novel bacterial cell extract-based DNA cloning method. Nucleic Acids Res 40(8):e55. doi:10.1093/nar/gkr1288

Zhou X, Cai S, Hong A, You Q, Yu P, Sheng N, Srivannavit O, Muranjan S, Rouillard JM, Xia Y, Zhang X, Xiang Q, Ganesh R, Zhu Q, Matejko A, Gulari E, Gao X (2004) Microfluidic PicoArray synthesis of oligodeoxynucleotides and simultaneous assembling of multiple DNA sequences. Nucleic Acids Res 32(18):5409–5417. doi:10.1093/nar/gkh879

Chapter 5
Tolerance of Microbial Biocatalysts to Feedstocks, Products, and Environmental Conditions

Mian Huang, George Peabody and Katy C. Kao

Abstract Bioreactor conditions and environmental stressors present during fermentation can negatively impact the productivity of industrial biocatalysts. Robustness of biocatalysts in fermentation conditions is thus important for the economical viability of bio-based production. Temperature, pH, and osmotic pressure inside the bioreactor are often not optimal for cell growth. Feedstocks (particularly sustainably sourced) and products (desired or side) often contain toxic components that further reduce biocatalyst performance. The physiological effects of many industrially relevant environmental stressors have been studied extensively. However, due to the complexity of cellular processes and the significant knowledge gap in genotype-phenotype relationships associated with these complex phenotypes, the rational engineering of robust biocatalysts is currently limited. Traditional strain developments rely on random approaches, and have been successful at generating more robust biocatalysts. Random approaches combined with new genomic technologies will start to fill the genotype-phenotype knowledge gap, making the rational engineering of robust biocatalysts for industrial applications more readily achievable. This chapter will focus on the common environmental stressors present in industrial fermentation; the stressors will be divided into three sections: feedstock toxicity, fermentation conditions, and product toxicity. Each section will describe the known mechanisms of toxicity associated with each stressor followed by examples of successful development of strains with enhanced tolerance, with a focus on the tools used, and discussions of the known molecular mechanisms associated with tolerance.

5.1 Introduction

Industrial fermentation conditions are often not optimal for production hosts. Inhibitory conditions can arise from process conditions (e.g., high temperature, low pH), type and amount of feedstock used and desired, and by-products formed.

M. Huang · G. Peabody · K.C. Kao (✉)
Department of Chemical Engineering, Texas A&M University, College Station, TX 77843-3122, USA
e-mail: kao.katy@tamu.edu

S. Van Dien (ed.), *Metabolic Engineering for Bioprocess Commercialization*,
DOI 10.1007/978-3-319-41966-4_5

The presence of inhibitors/inhibitory conditions can reduce the growth rate and productivity of the biocatalyst; thus production hosts robust to fermentation conditions are desirable. Unfortunately, producers do not necessarily have inherent high tolerance to fermentation conditions, and require strain optimizations for improved performance. In order to rationally engineer more robust microbial producers, some level of knowledge on the genetic determinants or molecular mechanisms involved is needed. However, due to the complex nature of how environmental conditions impact the cell and mechanisms by which biological systems can become more tolerant or resistant to these inhibitory conditions, our capacity for the rational engineering of more robust producers has been limited. However, with recent advances in genomics and high-throughput technologies, we are gaining regulatory and molecular insights on the cellular response and adaptation of biological systems at an unprecedented rate. Combined with random approaches for strain development, we can now use inverse engineering to shed light on the phenotype-genotype relationship in order to rationally engineer producers for enhanced tolerance to inhibitory production conditions.

These inhibitors can be divided into different categories: feedstock toxicity, environmental stressors, and product toxicity. This chapter will focus on these common environmental stressors present in industrial fermentation. Each set of inhibitors commonly present in industrial fermentation has been studied to determine the way in which it interferes with cellular activity so as to better approach industrial strain development. The large variety of feedstock, process conditions, and organic chemicals that biocatalysts are exposed to inevitably leads to an equally wide spectrum of cellular toxicities. The modes of toxicity are further complicated by the myriad availability of organisms that are and can be used, as different inhibitory compounds may elicit differential inhibitory effects on different strains. For example, although some mechanisms of adaptation towards inhibitory products are common to multiple organisms, each organism may also possess unique tolerance mechanisms. The wide range of responses to toxic conditions leads to a complex and vast space to explore to investigate the basis of tolerant phenotypes.

The types of known cellular toxicity are broad. Stressors often interrupt cellular function and productivity. Some modes of cellular toxicity are shared between the three common categories of environmental stressors, and include: destabilization of membrane properties, perturbation of the stability and function of enzymes, disruption of cellular functions through oxidation and other damaging reactions, and disrupted cellular homeostasis. Table 5.1 outlines the common modes of toxicity and Table 5.2 summarizes known mechanisms of tolerance, respectively, with example inhibitors.

To illustrate our current understanding of molecular mechanisms associated with enhanced robustness in industrially relevant conditions, we break down the variety of conditions into the three most common categories: feedstock toxicity, environmental stressors, and product toxicity. Each section will describe the known modes of toxicity associated with each stressor followed by discussions of the known molecular mechanisms associated with tolerance where applicable; otherwise, case

Table 5.1 Common inhibitory mechanisms associated with environmental stressors

Cytotoxicity		Inhibitors in feedstock	Fermentation conditions	Toxic products
Membrane Perturbations	Integrity (e.g. rigidity, fluidization)	Phenolics	Heat stress	Ethanol, butanol, isoprenoids, long-chain alcohols, fatty acids
	Perturbations in intracellular pH homeostasis	Weak organic acids, phenolics	Heat stress, low pH	Butanol, organic acids
	Perturbations in proton motive force	Weak organic acids, phenolics	Heat stress, low pH	Ethanol, organic acids
Enzyme denaturation		Furaldehydes, weak organic acids	Heat stress, low pH, osmotic stress	Ethanol, butanol
Undesirable Reactions	Reactive oxygen species formation	Furaldehydes	Heat stress	
	Interacting with nucleic acids	Furaldehydes, weak organic acids	Low pH	Organic acids
Disrupted cellular homeostasis	Imbalanced cellular redox state	Furaldehydes	NA	Isoprenoids, organic acids
	Disturbance of cell contents (e.g. ions, metabolites)	Salts, metals	Heat stress, osmotic stress	NA

NA not available

Table 5.2 Some known mechanisms associated with increased tolerance to cellular perturbations associated with environmental stressors

Tolerance/adaptation mechanisms		Inhibitors in feedstock	Fermentation conditions	Toxic products
Chaperones/stress response		Weak acids	Heat stress, osmotic stress, low pH	Ethanol, butanol, fatty acids, organic acids
Detoxification reactions/metabolization of toxic chemicals		Furaldehydes, weak acids	Low pH	Butanol, organic acids
Efflux pumps/active transport		Ionic liquids	Osmotic stress	Butanol, isoprenoids, long-chain alcohol, fatty acids
Alter membrane/physiological characteristics			Osmotic stress	Ethanol, butanol, long-chain alcohol
Other	Balance the cellular redox state	Furaldehydes	NA	NA
	Accumulation of compatible osmolytes (e.g. trehalose, glycerol)	Salts and high sugars	Heat stress, osmotic stress	NA

NA not available

studies of tools used for the successful development of strains with enhanced tolerance will be described. Finally, the chapter will conclude with a summary of how these tools can be applied to other inhibitors of interest.

5.2 Feedstock Toxicity

5.2.1 Lignocellulosic Biomass

The wide range of biomass feedstocks includes agricultural crops (e.g., corn and soybeans), food processing wastes, animal products (e.g., fats and manures), algae lipids, lignocellulosic biomass (e.g., sugarcane bagasse, wood, and straw), and others. Lignocellulosic biomass is relatively more abundant and not used as food for human consumption, which makes it one of the most promising raw materials for the biomass-based chemical and energy industry.

Lignocellulosic biomass is composed of three major types of polymers: cellulose, hemicellulose, and lignin. Cellulose is a linear highly ordered polymer of cellobiose; in cellulosic materials, the hydrogen bonds are formed between the hydroxyl groups of the glycosyl units from adjacent cellulose chains, holding these chains together and forming the fibrils with high tensile strength. In addition, the cellulosic materials have crystalline regions separated by amorphous regions. The crystalline regions are resistant to chemical or enzymatic hydrolysis due to their highly ordered structures, while the amorphous regions are sensitive to these attacks. Unlike cellulose, hemicellulose is a short, highly branched heteropolymer of complex monomeric units, including a variety of pentose and hexose, and has a loose and hydrophilic structure, which plays a role in holding the cellulose and lignin together. Lignin is a type of aromatic polymer, which consists of phenylpropanoid units derived from the corresponding p-hydroxycinnamyl alcohols. Unlike hemicellulose, lignin is hydrophobic and resistant to the chemical and biological degradation. In lignocellulosic materials, the lignin and hemicellulose form an amorphous matrix where cellulose fibrils are embedded and protected against hydrolysis. Thus, the utilization of the lignocellulosic biomass as feedstock for production of commodity chemicals and fuels requires energy-intensive pretreatments to break the matrix and release the carbohydrate polymers (cellulose and hemicellulose) that can be further hydrolyzed to produce fermentable sugars.

5.2.2 Known Mechanisms of Toxicity and Methods by Which They Were Determined

Depending on the nature and chemical composition of the lignocellulosic materials, different type of pretreatments may be used. Steam explosion and dilute acid pretreatments are relatively effective and inexpensive, thus are commonly used (Saha 2003). These thermo-chemical pretreatments are usually carried out under harsh conditions, such as high temperature and low pH, and lead to the formation of various inhibitory compounds in the resulting lignocellulosic hydrolysates. These compounds include degradation by-products of cellulose, hemicellulose, and lignin, and have negative effects on the fermenting microorganisms and the production of desired products (Parawira and Tekere 2011). Recently, pretreatment of lignocellulosic biomass using ionic liquids (ILs) emerged as a promising new approach for more efficient recovery of sugar, and low energy consumption (Portillo and Saadeddin 2014; Quijano et al. 2010). While the pretreatment does not generate the typical toxic inhibitory degradation products, IL itself is reported to be toxic to many bacteria, yeast, and fungi (Portillo and Saadeddin 2014; Quijano et al. 2010). The type and amount of biomass and the pretreatment method used contribute to the nature and concentration of the inhibitory compounds in the resulting lignocellulosic hydrolysates (Parawira and Tekere 2011). These inhibitory compounds can be classified into four major categories, furaldehydes, weak acids, phenolics, and ILs.

5.2.2.1 Furaldehydes

2-furaldehyde (furfural) and 5-hydroxymethyl-2-furaldehyde (HMF) are degradation by-products of pentose and hexose, respectively, and are common inhibitory furaldehydes found in lignocellulosic hydrolysates. These furaldehydes have been shown to negatively impact the growth of fermenting microorganisms, such as *Saccharomyces cerevisiae* and *Escherichia coli*, and reduce the yield and productivity of desired products, such as ethanol [reviewed in (Jonsson et al. 2013)]. Possible mechanisms underlying the negative effects on cell growth by the furaldehydes include reduced enzymatic and biological activities, DNA damage, inhibition in protein and RNA synthesis [reviewed in (Parawira and Tekere 2011)], induced oxidative stress (Allen et al. 2010; Kim and Hahn 2013), and imbalanced cellular redox state due to the depletion of NAD(P)H during the detoxification process (Ask et al. 2013a). In addition, furaldehydes with different chemical structures may have different inhibitory mechanisms; for example, furfural inhibits glycolytic enzymes and aldehyde dehydrogenase activities, leading to the intracellular accumulation of acetaldehyde, which can cause an extended lag phase in the growth of *S. cerevisiae* by inactivation of the cell replication [reviewed in (Parawira and Tekere 2011)]. Interestingly, the impact of furfural on ethanol production by *S. cerevisiae* is complex and dependent on its concentration and type of carbon source used in the fermentation; at low concentrations, furfural was found to stimulate the production of ethanol in xylose fermentation by a recombinant *S. cerevisiae* strain [reviewed in (Jonsson et al. 2013)]. During the ethanolic fermentation of xylose by a recombinant yeast strain, the formation of xylitol by xylose reductase (XR) generates less NAD^+ than that required by xylitol dehydrogenase (XDH) to oxidize all produced xylitol to xylulose, which results in xylitol excretion and lowers the ethanol yield. The addition of low concentration of furfural increases supply of NAD^+ during the reduction of furfural to furfuryl alcohol in yeast, which compensates for the shortage of NAD^+ in the xylose conversion; as a result, it decreases the xylitol excretion and increases the ethanol yield (Jonsson et al. 2013).

5.2.2.2 Weak Acids

There are three commonly found weak acids in lignocellulosic hydrolysates: acetic, formic, and levulinic acid. Acetic acid is formed primarily by the de-acetylation of hemicellulose, while the other two aliphatic acids are derived from the acid-catalyzed thermo-chemical degradation of polysaccharides (Jonsson et al. 2013). Currently, there are two major mechanisms proposed to explain the inhibitory effects of the weak acids: uncoupling and intracellular anion accumulation [reviewed in (Russell 1992)]. The uncoupling theory assumes that weak acids have similar effects on the proton motive force and the electrical gradient across the cell membrane as an uncoupler, which inhibits the synthesis of ATP by disrupting the coupling between electron transport and phosphorylation reactions. The uncoupler can diffuse across the cell membrane and dissociate inside the cells due to the higher cytosolic pH, which releases protons and decreases the intracellular pH. The anion of the dissociated uncoupler is then driven outside the cells by the electrical gradient and gets protonated due to the lower extracellular pH to complete the cycle of proton translocation. In order to maintain a constant intracellular pH, cells have to pump out the protons brought in by the uncoupler at the expense of ATP that could otherwise be used for biomass formation, thereby inhibiting cellular growth. Moreover, the acidified intracellular environment could affect the integrity of purine bases, and result in mutations and damage to cellular DNA and RNA (Warnecke and Gill 2005). Although the weak acids are lipophilic and can pass through the cell membrane easily, their polar anionic forms are lipophobic and thus cannot diffuse freely across the cell membrane. Thus, another theory based on the permeability of cell membrane to the different forms (dissociated and undissociated) of the weak acids was developed [reviewed in (Russell 1992)], which states that the intracellular accumulation of high and potentially toxic concentrations of anionic forms of the weak acids is the major cause of the cell growth inhibition. The accumulation of large amount of anions dissociated from the weak acids inside a cell may also disrupt its intracellular anion pool, resulting in the transport of excessive potassium into the cell and thus increasing osmotic pressure. In order to maintain a constant osmotic pressure, more glutamate will be pumped out of the cell, which disrupts the osmolarity of the cytoplasm and lowers growth potential and viability [reviewed in (Warnecke and Gill 2005)]. Finally, weak acids have also been shown to inhibit cell growth by reducing the uptake of certain aromatic amino acids from the culture medium through inhibiting amino acid permease (Bauer et al. 2003). Similar to furfural toxicity, some weak acids, including acetic, formic, and levulinic acid, have strong inhibitory effects on cell growth at high concentrations, whereas lower concentrations are able to enhance product yield (Larsson et al. 1999).

5.2.2.3 Phenolics

A large number of different phenolic compounds can be released during the breakdown of lignin and degradation of sugars during thermo-chemical pretreatments, and cause decreased biomass yield, growth rate, ethanol yield and productivity [reviewed in (Almeida et al. 2007)]. The most common phenolics found in pretreated lignocellulosic materials are 4-hydroxybenzaldehyde, 4-hydroxybenzoic acid, vanillin, dihydroconiferyl alcohol, coniferyl aldehyde, syringaldehyde, and syringic acid (Klinke et al. 2004). Depending on the molecular weight (MW), substituent position in the benzene ring, and hydrophobicity of the functional group, phenolics may have either strong or weak inhibitory effects on the microorganisms. Generally, phenolics with lower MW and more hydrophobic functional group, such as aldehydes and ketones, may have stronger inhibitory effects on *S. cerevisiae* [reviewed in (Almeida et al. 2007)]. In addition, hydroxyl and methoxyl groups in *ortho* and *meta* positions increase the toxicity of vanillins when compared to their substituents in *meta* and *para* positions or vice versa [reviewed in (Almeida et al. 2007)]. Due to the structural diversity and the lack of accurate methods to characterize and quantify compounds of this group, the mechanisms of toxicity of phenolics on microbial systems are largely unknown. One possible mechanism of toxicity is that phenolics may interfere with the normal functioning of cell membranes by negatively impacting the integrity and altering the lipid-to-protein ratios of membranes [reviewed in (Almeida et al. 2007)]. Furthermore, some acidic groups on phenolic compounds can act as uncouplers, and can inhibit the oxidative phosphorylation in mitochondria by altering the electrochemical gradient across its membrane (Terada 1990).

5.2.2.4 Ionic Liquids

Ionic liquids are organic salts typically composed of large organic cations, such as 1-butyl-3-methylimidazolium cation ([Bmim]), and small inorganic anions, such as chloride anion ([Cl^-]), that exist as liquids below a threshold temperature (around 100 °C). These organic salts are used in the pretreatment of lignocellulose biomass because they can dissolve a large amount of cellulose at considerably more mild conditions compared with other pretreatment methods, and the dissolved polysaccharides can be easily regenerated by addition of an anti-solvent, such as acetone (Vancov et al. 2012). The ability of ILs to dissolve cellulose varies and is primarily controlled by the anion components. ILs with chloride anion were found to have high efficiency in cellulose dissolution (up to ~25 wt% of cellulose) (Vancov et al. 2012). The toxic effects of ILs to a variety of microorganisms have been studied in terms of cell growth and viability, product formation, and substrate uptake. However, due to the lack of standardized approaches, the results from different studies are difficult to compare and sometimes lead to contradictory observations. Generally, ILs are toxic to microorganisms at or above a concentration of ~10 % (v/v) (Quijano et al. 2010). The toxicity of ILs also seems to be mainly determined by the cation components; cations with longer alkyl chains and more substituted groups tend to be more toxic (Quijano et al. 2010). However, little is known about the toxic mechanisms and cellular response of microbial systems to ILs at the molecular level, which remains to be addressed in future studies.

5.2.3 Efforts to Enhance Tolerance

The presence of various inhibitory compounds in lignocellulosic hydrolysates negatively impacts the growth and subsequent fermentation of industrial producers, resulting in extended lag phase during fermentation, poor biomass yield, and low productivity and yield of desired products. Although several efficient chemical, enzymatic, and biological detoxification approaches have been successfully developed to reduce the inhibitory effects of lignocellulosic hydrolysates, the use of microbial producers with enhanced tolerance is desirable as minimal or no detoxification may be required, which reduces process costs and avoids loss of fermentable sugars during the detoxification treatments [reviewed in (Parawira and Tekere 2011)]. The major strategies to engineer tolerant strains include metabolic engineering and evolutionary engineering; both have achieved modest successes thus far.

5.2.3.1 Metabolic Engineering

The rational engineering of microorganisms to be more tolerant to inhibitors present in lignocellulosic hydrolysates requires *a priori* knowledge on the target genes and/or pathways involved in the desired phenotypes. Several high-throughput approaches, including transcriptome analysis, protein-based analysis, and genome library-based screening, have been successfully applied to identify target genes for subsequent metabolic engineering of strains for enhanced tolerance to various inhibitors. Several excellent review papers have summarized these aforementioned strategies (Ling et al. 2014; Parawira and Tekere 2011). Here, we briefly introduce some relevant examples from these review papers that represent recent advances in tolerance engineering of microbes toward various lignocellulose-derived inhibitors.

In one case of using transcriptomics to identify target genes, Petersson and coworkers identified a number of genes that are involved in the HMF reduction, including *ADH6*, *ADH7*, and *SFA1*, by comparing the transcript profiles of a control strain and one strain of *S. cerevisiae* with enhanced tolerance to lignocellulosic hydrolysates, both in the presence and absence of exogenous HMF (Petersson et al. 2006). Furthermore, they confirmed the role of the *ADH6* in the HMF reduction both *in vitro* and *in vivo* and that its HMF reduction capacity is NADPH-dependent. In another study, researchers identified a mutated *ADH1* gene to be involved in NADH-dependent HMF reduction in an industrial strain TMB3000 by extracting and fractionating the cytosolic proteins from the strain (Laadan et al. 2008). Three missense mutations were identified in the *ADH1* gene, which the authors presumed to be responsible for the unusual substrate specificity. More importantly, overexpression of the mutated *ADH1* resulted in the reduction of growth lag phase in the presence of HMF, increased *in vivo* HMF reduction rate, and higher ethanol production rates. The HMF reduction activity may lead to decreased NAD(P)H pool, cofactors of the detoxification enzymes, and other critical biosynthetic enzymes, which may have different physiological impacts on different microorganisms. In *S. cerevisiae*, the increased consumption of NAD(P)H for the reduction of HMF could be compensated by increased production of acetate and the resulting induced stress response, which facilitates the regeneration of depleted cofactors (Almeida et al. 2008). In *E. coli*, the HMF reduction appears to negatively impact cell growth, as a result of competition for cofactors between the detoxification enzymes (e.g., yqhD and dkgA) and critical biosynthetic enzymes (e.g., cysJ, thrA, and dapB) (Miller et al. 2009). Thus, the associated physiological effects need to be taken into account when designing strain engineering strategies to increase the capacity of HMF reduction in target microorganisms for better tolerance. Using a different approach, Gorsich et al. screened a *S. cerevisiae* gene disruption library for genetic determinants associated with furfural tolerance; they identified 62 genes from a variety of cellular processes from their screen, suggesting that furfural tolerance is a complex phenotype (Gorsich et al. 2006). Four of the 62 mutants contain mutations in genes that function in the pentose phosphate pathway (PPP), which led to their finding that *ZWF1* and PPP contribute to furfural tolerance by balancing the levels of NADPH consumed in the furfural detoxifications and cell repair processes in response to the furfural-induced damage.

Based on the known toxicity of furfural and HMF, metabolic engineering efforts focused on the oxidative stress response and cofactor-balancing have been used to enhance strain tolerance to the furaldehydes. In *S. cerevisiae*, overexpression of the transcription factor Yap1 (involved in oxidative stress response) and its two downstream targets CTA1 and CTT1 (catalases) were shown to enhance yeast tolerance to both HMF and furfural (Kim and Hahn 2013). Furthermore, engineering the cells to overexpress GSH1 and GLR1, genes involved in the glutathione (GSH) metabolism, increased the robustness of the *S. cerevisiae* to inhibitors present in the spruce hydrolysates and enhanced the ethanol titer and yield in a simultaneous saccharification and fermentation process (Ask et al. 2013b). Research also showed that increasing GSH level by overexpression of GSH1 and GLR1 or by the exogenous addition of GSH to the culture medium enhanced the tolerance of the *S. cerevisiae* to furfural (Kim and Hahn 2013).

5.2.3.2 Evolutionary Engineering

In addition to rational engineering by genetic manipulation, evolutionary engineering is an alternative strategy used in development of strains with enhanced tolerance to various inhibitors. Evolutionary engineering does not require prior knowledge on the details of the action of inhibitors and their complex interaction with the biological systems, which is extremely useful for complex or understudied phenotypes. Demeke et al. engineered an industrial bioethanol producing yeast strain for xylose utilization, then performed random mutagenesis using ethyl methanesulfonate (EMS), followed by genome shuffling (via mating/sporulation) and selection in lignocellulosic hydrolysate, and multiple rounds of evolutionary engineering (Demeke et al. 2013); the resulting isolated strain is able to utilize both D-glucose and D-xylose efficiently and exhibits high ethanol yield when growing in various lignocellulosic hydrolysates. In another study, a recombinant xylose utilizing *S. cerevisiae* strain was evolved for enhanced tolerance to inhibitors present in spruce hydrolysate via two different evolutionary strategies: the adaptation in repetitive batch cultures with a cocktail of 12 inhibitors (synthetic hydrolysates) and adaptation in chemostats using spruce hydrolysate (Koppram et al. 2012). Phenotypic characterization of the evolved strains suggests that different evolutionary strategies may select for different physiological properties. Evolution in repetitive batch cultures favored the selection of higher maximum specific growth rate, whereas evolution in chemostat improved the specific conversion rate of the inhibitors. This difference emphasized the importance of choosing an appropriate strategy for engineering target phenotypic traits. The choice of strategy will be case-dependent. For example, serial batch transfer will also select for mutants with reduced growth lags in addition to increased specific growth rates. However, the evolved strains generated from both strategies showed improved fermentation performance in spruce hydrolysates under anaerobic conditions. In another work, Almario et al. evolved *S. cerevisiae* via serial batch transfers for enhanced tolerance to corn stover hydrolysates and used an adaptive laboratory evolution method called visualizing evolution in real-time (VERT) to track the development of hydrolysates tolerance and to isolate adaptive clones for characterizations and molecular analyses (Almario et al. 2013). The isolated mutants from their evolution study showed improved growth rates in the hydrolysates with the largest improvement up to 57 % over the parental strain.

During evolutionary engineering, the selective pressure (inhibitor concentration) can be ramped-up to generate more tolerant strains. If no measurement phenotypes are available to assess improvement, then trial-and-error is often the strategy used for ramp-up. Fortunately, if serial batch transfer was used with inhibitor tolerance, increased tolerance generally results in increased specific growth rates, increased final biomass concentrations, and/or reduction in the lag phase. These growth-related parameters can be readily measured during the course of evolution and used as feedback for increasing the selective pressure. If continuous cultures are used, increased tolerance may lead to increased biomass concentration in the bioreactor. However, it is important to note that these measurements are population-averages, and thus may not be sensitive enough to detect the presence of beneficial mutants in the population. VERT is a more sensitive method, that uses changes in relative frequencies of different colored subpopulations as an indication of the rise and expansion of adaptive mutants (Reyes et al. 2012). Indeed, the use of this method was successfully applied to rationally ramp-up the selective pressure in evolving hydrolysates tolerance in *S. cerevisiae* (Almario et al. 2013).

5.3 Fermentation Conditions

5.3.1 *Temperature, PH, and Osmotic Stress*

In addition to the stressors present in the feedstock, industrial microorganisms are routinely exposed to non-optimal environments during fermentation. Some challenging conditions encountered include high thermal stress (e.g., during simultaneous saccharification and fermentation), low pH (e.g., during production of organic acids), high osmotic stress (e.g., during fermentation with high substrate concentration). Production performance of industrial strains is partly dependent on their ability to adapt to these challenges; thus, understanding cellular responses and tolerance mechanisms to these stressors is important. Here, we focus on three critical stresses commonly encountered by microorganisms during fermentation (heat stress, low pH, and osmotic stress), and summarize the current knowledge on their physiological effects, stress responses and adaptation, and engineering tools to enhance stress tolerance.

Microorganisms are exposed to ever-changing conditions in their natural environments and thus are able to adjust their metabolisms in response to moderate fluctuations in environmental conditions. However, when a given condition, such as pH, falls outside its normal growth range for the particular strain, it will negatively impact cellular growth, resulting in reduced growth rate and/or extended lag phase (Cheroutre-Vialette et al. 1998). While cellular responses to the stressors may share some commonality, they are generally unique to the type of stress. The cellular damage resulting from heat stress includes loss of membrane integrity, accumulation of intracellular oxidants, damage to protein and enzymes, internal acidification, and metabolic disequilibria (Attfield 1997; Lindquist 1992), which may ultimately result in cell death. Low external pH caused by the presence of either weak organic acids or strong inorganic acids in the culture medium is another important type of stress encountered by industrial microorganisms. The low external pH causes internal acidification. The mechanisms underlying the internal acidification may differ and depend on the type of acids causing the stress. Here, we focus our discussion on stress resulting from strong inorganic acids and refer the reader to previous sections (Sect. 5.2.2.2) for discussion on weak organic acids. Unlike weak organic acids, strong inorganic acids do not readily diffuse through cell membranes, but cause internal acidification by increasing proton influx through large pH gradient spanning the cytoplasmic membrane (Beales 2004). In addition, strong acids are able to inactivate proteins and enzymes present on the cell surface, including those involved in the uptake of essential ions and nutrients (Booth 1989). These pH-associated effects may lead to reduced growth rate and extended growth lag (Cheroutre-Vialette et al. 1998). Osmotic stress usually arises due to a sudden increased or decreased concentration of solutes present in the surroundings. Cells that are exposed to a higher external osmolarity experience a hyperosmotic stress. In *S. cerevisiae*, hyperosmotic stress causes a reduction of cell volume, loss of turgor, growth inhibition, disturbance of metabolite concentration, and reduced fermentation activity (Attfield 1997). In *E. coli*, osmotic stress was reported to drastically inhibit active transport of carbohydrates (Roth et al. 1985), DNA replication and cellular division (Meury 1988).

5.3.2 Known Mechanisms for Each Stressor and Methods by Which They Were Determined

In response to environmental stresses, various cellular processes are involved in coping with the stress-associated cell damage and promote cell survival, including growth control, cell sensing, signal transduction, transcription, and post-translational control [reviewed in (Attfield 1997)]. These stress responses have been extensively characterized in model organisms such as *S. cerevisiae* and *E. coli*. However, we are still far from fully understanding the molecular basis underlying these stress responses, especially the mechanisms by which cells sense a given environmental stress and transmit the signal to activate corresponding transcription factors. In this section, we summarize current understandings of the cellular responses to selected environmental stresses.

5.3.2.1 Heat Stress

Heat stress responses (HSR) have been extensively studied in *S. cerevisiae*; several major transcription factors are known to be activated in response to elevated temperature. These major players of HSR include heat shock transcription factor 1 (*HSF1*) and transcription factors encoded by *MSN2* gene and its close homolog *MSN4* gene. The Hsf1p is activated by hyper-phosphorylation in response to heat stress and positively regulates the expression of genes containing the heat shock elements (HSEs) in their promoters. A study using a loss-of-function mutant, *hsf1-R206S/F256S*, identified 72 genes whose heat shock induction depends on the functional Hsf1p (Yamamoto et al. 2005). Unlike Hsf1p, Msn2/4p is activated in response to a variety of stresses, including carbon source starvation, heat shock, severe osmotic and oxidative stresses, and their activation induces expression of genes containing the stress response elements (STREs) in their promoters (Martinez-Pastor et al. 1996). Typical genes involved in the more general STRE-dependent response are diverse in their functions, such as UBI4 (ubiquitin); DDR2 (DNA damage repair); HSP104, 26, and 12; CTT1 (catalase T); TPS1 (trehalose-6-phosphate synthase) and TPS2 (trehalose phosphate phosphatase); and numerous others (Mager and De Kruijff 1995; Thevelein 1994). The mechanisms whereby yeast senses thermal stress and transmits signals to heat-induced transcription factors are not well characterized. In terms of activation of *HSF1*, current studies support a model where accumulated misfolded proteins, resulting from thermal damage, can effectively compete for binding to repressors of Hsf1p, such as

Hsp70, Hsp90, and their cofactors, so that the Hsf1p is released and activated (Morano et al. 2012). Activation of these transcription factors and other unknown transcriptional networks in response to heat stress leads to the accumulation of many proteins and molecules involved in cell survival and repair of cellular damage, including heat shock proteins (HSPs), trehalose, and glycogen. HSPs function as protein chaperones and their cofactors, protecting cells from the thermal damage by either refolding or degrading misfolded proteins (Morano et al. 1998). Trehalose acts as a powerful stabilizer of proteins to protect them from thermal denaturing and aggregation (Hottiger et al. 1994; Kaushik and Bhat 2003). In addition, accumulation of glycogen is suggested to play a role in thermotolerance, possibly as a reserve carbohydrate (Parrou et al. 1997; Unnikrishnan et al. 2003).

5.3.2.2 Osmotic Stress

Molecular basis of the response to osmotic stress, especially to hyperosmotic stress in yeast, is relatively well characterized (Hohmann 2002). In response to hyperosmotic stress, yeast induces expression of genes involved in synthesis of compatible solutes, efflux of cations, and general stress response. The key signaling pathway to osmotic stress is the high osmolarity glycerol (HOG) pathway, which is activated by salts and other osmolytes to protect cells from associated cell damage (Chen and Thorner 2007). HOG activation involves two upstream osmosensors, Sln1 and Sho1, which are transmembrane proteins, and two protein kinases, Pbs2 and Hog1, which trigger osmoresponsive genes and other STRE-dependent general stress responsive genes once activated [reviewed in (Saito and Posas 2012)]. A crucial change in cell metabolism after activation of HOG pathway is the hyper-production and hyper-accumulation of glycerol, which functions as a compatible solute to protect cells by balancing the internal and external osmolarities [reviewed in (Saito and Posas 2012)]. When exposed to hyperosmolarity exerted by NaCl, the ENA1 gene, which encodes a Na^+, Li^+-ATPase, is overexpressed in a calcineurin-dependent manner in yeast, leading to the efflux of Na^+ and Li^+ (Mendoza et al. 1996). Other mechanisms employed by yeast to respond and adapt to hyperosmotic stress include change in the membrane structure, vacuolar functioning, and accumulation of trehalose level [reviewed in (Gibson et al. 2007; Hohmann 2002)].

Osmoadaptive mechanisms of bacteria from diverse phylogenetic groups have been extensively studied. The most common theme of osmotic response from these studies is that cells accumulate or release various solutes in response to fluctuations in environmental osmolality (Wood 2015). These solutes include inorganic ions, such as K^+, and small organic molecules, such as trehalose, glutamate, and glycine betaine, that do not significantly perturb cellular functions even at high intracellular concentrations (Altendorf et al. 2009). Under hyperosmotic stress, cells become dehydrated, leading to perturbation in various cellular functions and reduced population growth rates (Wood 1999; Cayley and Record 2004; Altendorf et al. 2009). The accumulation of the osmoregulatory solutes can rehydrate the stressed cells and depending on the nature of the solute, can restore cellular growth to varying extents

(Cayley and Record 2004). K^+ glutamate partially rehydrates the cells and inhibits the binding of protein to nucleic acids, which cannot restore growth to the pre-stress level. In contrast, organic osmoregulatory solutes are able to restore cell hydration and growth more effectively. In response to elevated osmolality, cells can accumulate osmoregulatory solutes by two routes, uptake from the environment and de novo biosynthesis. These processes are well characterized in *E. coli* and multiple enzymes, transporters, and channels were involved (Altendorf et al. 2009). The uptake of K^+ in *E. coli* is mediated by K^+-H^+ symporters Trk and a P-type ATPase K^+ transporter Kdp. The former are encoded by trkH and trkG and function in conjunction with gene products of trkA and sapD, while the latter contains subunits encoded by the kdpFABC operon. In terms of the uptake of organic solutes, there are four known transporters available in *E. coli*, including a major facilitator superfamily member ProP, an ABC transporter ProU, and two betaine-carnitine-choline family members BetT and BetU. In addition, enzymes involved in the biosynthesis of organic osmoprotectants, such as trehalose and glycine betaine, have also been identified. OtsA and OtsB are responsible for the synthesis of trehalose, while BetA and BetB are required for the production of glycine betaine from choline, whose uptake was mediated by BetT (Murdock et al. 2014). Most of the known osmoregulatory systems in bacteria are regulated transcriptionally and the best-studied cases include two-component systems (TCS), such as the KdpD/KdpE system (Kramer 2010; Altendorf et al. 2009). KdpD and KdpE are the histidine kinase and the response regulator of this TCS respectively. The KdpD/KdpE system responds to K^+ limitation and salt stress and targets the kdpFABCDE operon, which includes genes encoding the TCS and the aforementioned Kdp ATPase. There also exist systems that can be mediated at the level of protein activity, such as ProP in *E. coli* (Kramer 2010). ProP mediates the H^+-coupled uptake of small organic solutes, for example proline and glycine betaine, whose activity increases with increasing external osmolality, leading to faster uptake rate of solutes in response to osmotic upshift.

5.3.2.3 Low pH

Responses and tolerance to low pH stress are understudied in yeast and much remains unknown regarding pH sensing, signal transduction, transcription regulation, and other cellular responses to this stressor. Some of the responsive gene targets in yeast have been identified using transcriptional profiling; these include cell wall-related genes, genes involved in carbohydrate metabolism and redox metabolism, transcription factors, and STRE-dependent general stress responsive genes (Kapteyn et al. 2001). In bacteria, several systems contributing to cell survival in acidic environments have been characterized, including acid-tolerance response (ATR) system, which can function in minimal medium to protect cells in log phase or stationary phase from acid at pH levels as low as 3.0, and acid-resistance (AR) system, which protects cells in stationary phase to even lower pH values (<2.5) (Lin et al. 1995). ATR systems are present in a variety of

Gram-negative and Gram-positive bacterial species, and they can offer protection against lethal acid stress once induced by pre-exposure to a sublethal level of acid stress (Boot et al. 2002). The induction of the system involves a number of acid-shock proteins (Choi et al. 2000), which are involved in modification of membrane composition (Chang and Cronan 1999; Jordan et al. 1999; Yuk and Marshall 2004), internal pH homeostasis (Park et al. 1996; Richard and Foster 2004), and repair or protection of essential cellular components (Choi et al. 2000; Raja et al. 1991). Three distinct AR systems are currently known in *E. coli* and *Shigella flexneri* (*S. flexneri*), including an oxidative-dependent AR, an arginine-dependent AR, and a glutamate-dependent AR system (Lin et al. 1995). The oxidative-dependent system is induced in acidic environment in stationary phase cells, whose expression requires the alternative sigma factor RpoS and the cyclic AMP receptor protein CRP (Castanie-Cornet et al. 1999). In addition, the system is under catabolite repression. However, the detailed mechanism of how this oxidative-dependent AR system functions to protect cells from low pH remains elusive. The other two AR systems function in the presence of specific amino acids and protect cells from acid stress by consuming intracellular protons during decarboxylation reactions, which can be catalyzed by arginine decarboxylase (AdiA) and glutamate decarboxylase (GadA and GadB) (Castanie-Cornet et al. 1999), and changing the membrane potential to reduce proton permeability (Richard and Foster 2004).

5.3.3 Efforts to Enhance Tolerance

Similar to inhibitors present in feedstock, a variety of engineering tools have been used to improve the tolerance of microorganisms to various fermentation conditions of industrial relevance, including single gene manipulation, transcription factor-based engineering, whole genome shuffling (WGS), small RNA-based post-transcriptional regulation, and evolutionary engineering.

Manipulation of single genes either via overexpression or deletion is a useful technique in rational engineering for desired phenotypes. Cellular tolerance to stressors involves activation of many signaling and regulatory pathways, which makes strain tolerance engineering through direct manipulation at the molecular level very difficult. Though challenging, several prior efforts have successfully improved tolerance to fermentation conditions via single gene manipulation. Overexpression of *RSP5*, encoding an E3 ubiquitin ligase, was found to confer superior thermotolerance to *S. cerevisiae*, possibly by facilitating the removal of stress-damaged proteins by increasing the level of ubiquitination (Shahsavarani et al. 2012). In addition, overexpression of *TPS1*, encoding a trehalose-6-phosphate (T6P) synthase in yeast, results in enhanced ethanol production, from 37 to 60 g/L, and thermotolerance (improved critical growth temperature from 36 to 42 °C) (An et al. 2011). These examples suggest that some key components of the stress responses may be good targets to manipulate the stress tolerance.

Often times the cell is encountering multiple stressors at once. Thus rational strain improvement will require knowledge of the major stressor(s) involved. The question of how to diagnose which stressor is the cause of poor fermentation performance is a challenging one to address, due to the presence of multiple stressors and potential synergistic effects of combinations of stressors. Transcriptomic and/or proteomic analysis may be a quick way to identify the major stressor(s) based on the cellular response in fermentation conditions using existing biological knowledge. Alternatively, if the identities and concentrations of the potential stressors are known, then controlled experiments involving each individual and combinations of stressors can be performed to identify the most relevant stressors (or combination of stressors) for which enhanced tolerance is needed to improve performance. However, if the above strategies are not possible, then more global approaches, such as the ones described below, may be applied.

Transcription factor-based engineering has become an important class of tools for strain development; the manipulation of transcriptional regulators allows perturbing the expression of multiple genes simultaneously. Global transcription machinery engineering (gTME) was initially developed in bacteria and yeast and represents the first systemic tool of this class. Since the initial development, gTME has been successfully applied in *Lactobacillus plantarum* to select for mutants with enhanced lactic and inorganic acid tolerance by targeting the principal σ factor, encoded by *RpoD* (Klein-Marcuschamer and Stephanopoulos 2008). The same technique that targets two components of RNA polymerase II transcription factor D (TFIID), TATA-binding protein SPT15 and TATA-binding associated factor TAF25, of *S. cerevisiae* has been used to select mutants with enhanced tolerance to high concentrations of ethanol (6 % v/v) and glucose (120 g/L) (Alper et al. 2006). CRP is a well-known global regulator that controls the expression of over 400 genes in *E. coli* (Khankal et al. 2009), and was successfully targeted for the strain development of osmotolerant (Zhang et al. 2012) and acid tolerant mutants (Basak et al. 2014).

WGS has also been used to enhance *S. cerevisiae* tolerance to multi-stresses, including harsh fermentation conditions; where the DNA of similar but diversified cell populations is shuffled to recombine desirable genes or mutations. Shi et al. successfully improved the thermotolerance, ethanol tolerance, and ethanol productivity of an industrial yeast strain SM-3 using this technique (Shi et al. 2009). Their starting population was created by protoplast ultraviolet irradiation and then subjected to three rounds of recursive protoplast fusion; positive colonies were screened for growth in the presence of combinations of different temperatures and ethanol concentrations. The best performing strain, F34, was capable of growth in up to 55 °C, producing 9.95 % (w/v) ethanol, and tolerating 25 % (v/v) ethanol stress.

Small RNAs (sRNAs) are a group of small non-coding RNA molecules, which could modify the function of proteins and regulate gene expression by binding to target molecules (Vogel and Wagner 2007). The sRNAs-based technique in metabolic engineering is emerging and has been applied to improve biological systems for stress tolerance. A recent successful example of sRNA-based approach is the overexpression of DsrA, RprA, and ArcZ (individually and in combinations) in *E. coli* to significantly enhance the tolerance to low pH stress (pH 2.5) (Gaida et al. 2013). Simultaneous overexpression of all three sRNAs resulted in an 8500-fold increase in survival in acid conditions compared with the control strain. In addition to the enhanced tolerance to low pH, overexpression of the three sRNAs also conferred significant protection to carboxylic acid and oxidative stress (Gaida et al. 2013). The proposed mechanism by which the overexpression of the three sRNAs may protect cells from the stresses is by altering the mRNA levels of two critical regulators involved in stress tolerance, RpoS and H-NS, and their downstream gene targets. Unlike tools based on gene overexpression, sRNA-based tools generally impose less metabolic burden on cells, and thus may be more effective in strain engineering.

Evolutionary engineering, as described earlier, was recently applied to improve the osmotolerance (NaCl) in *E. coli* (Winkler et al. 2014). Compared to the unevolved parental strains, isolates from the evolution experiments had significantly improved growth rate under the osmotic stress and could grow in up to 0.8 M (47 g/L) NaCl in minimum media, a concentration that completely inhibits the parental strain.

5.4 Product Toxicity

5.4.1 Biosynthesis Targets and Their Potential Applications

There are a myriad of targets for bio-based production, with applications in areas such as biofuels, platform chemicals, and therapeutics, as the range of natural and non-natural compounds that can be produced using microbial systems is incredibly vast. The classes of desired compounds include isoprenoids, fatty acid derivatives (including long-chain alcohols), and less complex chemicals with more direct synthesis such as short-chain alcohols and organic acids (e.g., acetic acid, butanol, and ethanol). Isoprenoids are promising targets for bio-based production due to their structural complexity, which renders typical chemical synthesis difficult. This class of compounds has potential applications in food, pharmaceuticals, cosmetics, and energy industries. Isopentenyl-diphosphate (IPP) and the downstream product dimethylallyl pyrophosphate (DMAPP) are the central precursors of isoprenoid compounds and can be made via two different routes; these form the foundation for the biosynthesis of downstream isoprenoid compounds. Notable isoprenoid products include: lactone artemisinin, cyclic terpenes, pinene, geraniol, alpha-pinene, limonene, and farnesyl hexanoate (Peralta-Yahya et al. 2012); several of these compounds, including the intermediate IPP, are known to be toxic to microbial systems. Another important class of biosynthetic compounds is fatty acids and their derivatives. Fatty acid-derived products are typically synthesized from malonyl-CoA, where the malonyl-CoA base is extended by sequential decarboxylative condensations of additional malonyl-CoA building blocks. The products that have been made via this pathway vary depending on the final modification of the fatty acid produced, and include FAEEs (fatty acid ethyl esters), FAMEs (fatty acid methyl ester), alkanes and alkenes (Peralta-Yahya et al. 2012). Many fatty acid-derived compounds are biofuels, for example FAMEs and FAEES as biodiesel, and alkanes as gasoline (Peralta-Yahya et al. 2012). In addition, as suggested by the list of value-added chemicals from biomass released by the Department of Energy, several organic acids, such as succinic acid, itaconic acid, and levulinic acid, are of industrial significance as platform chemicals and preservatives. However, organic acids are generally toxic to microbial systems. Alcohols, both short and long chain, are currently highly sought after for commercialization using biological-routes due to their potential use as biofuels and platform chemicals. The most common short-chain alcohols targeted for bio-based production are n-butanol, isobutanol, and ethanol. N-butanol is naturally produced by *Clostridium* spp. as part of the ABE (acetone, butanol, and ethanol) fermentation process, and has been studied for decades. The biosynthesis of isobutanol has more recently been targeted. Short-chain alcohols have well-studied toxicities, whereas long-chain alcohols are less studied in microbial systems.

5.4.2 *Modes of Toxicity and Known Mechanisms of Tolerance*

Since the tools used to characterize cellular toxicity and adaptation to toxic products are largely the same as those used to study inhibitory feedstock and fermentation conditions, we will focus on the current known modes of toxicity for different classes of bio-based products. Instead of focusing on the tools used, where progress has been made, we will discuss the current understanding of mechanisms to alleviate their toxicities.

5.4.2.1 Isoprenoids/Terpenoids

Due to their structural complexity, isoprenoids are generally excellent targets for bio-based production. However, many have poorly understood biological roles and therefore low level of comprehension of their interaction with cellular components, especially when present in high concentrations. The more characterized isoprenoid compounds have displayed a range of toxicity when exogenously added, though current production levels may be too low to trigger extensive toxicity. It is theorized that isoprenoids localize to the cell membrane due to their lipophilic nature. By intercalating or otherwise inserting themselves in the membrane, these compounds likely cause problems associated with membrane fluidity such as loss of chemostatic control (Sikkema et al. 1995). Some specific isoprenoids have been studied more extensively, such as pinene and limonene. Alpha-Pinene has been shown to inhibit respiration of *S. cerivisiae* within the mitochondria, presumably by acting on the cytochrome B protein, an integral part of respiration. Cyclic terpenes have also been observed to accumulate in the membrane, though any associated toxicity has yet to be studied in detail (Sikkema et al. 1995). Beta-Pinene has also been shown to affect membrane function and disrupt ion (specifically potassium and proton) translocation (Sikkema et al. 1995), thereby inhibiting respiration. Limonene has been observed to similarly inhibit respiration in mitochondria as well. Lactone artemisinin is a strong antimalarial drug, however the toxicity of the compound appears to be unique to the protozoan malaria, where it kills asexual stages of the parasite by inhibiting a calcium ATPase. It can therefore be produced in relatively high concentrations by various other microbial organisms with limited toxicity issues (Gershenzon and Dudareva 2007).

Existing mechanisms of tolerance toward isoprenoids are limited. The only well-documented success is in utilizing active transport to increase tolerance. Several groups have reported expressing proteins known for efflux of drugs and lipids that result in increased tolerance to a variety of isoprenoids in *E. coli* such as geraniol, alpha-pinene, limonene, and farnesyl hexanoate (Dunlop 2011; Jin et al. 2014). As the host productivity for isoprenoids is optimized, both characterization of the toxic effect of isoprenoids and mechanisms to alleviate their toxicity will become more critical toward industrial applications.

5.4.2.2 Fatty Acids

Fatty acids and fatty acid-derived products have varying microbial toxicity, where certain products are essentially harmless and others are highly toxic. FAEEs have been observed to not inhibit growth of *E. coli* at concentrations up to 100 g/L (Steen et al. 2010), and presumably FAMEs exhibit similar lack of toxicity. Alkanes and alkenes can be produced from fatty acids, have high level of hydrophobicity, and thus can disrupt membrane function and structure. They have been shown to impair membrane containment of ions, metabolites, and lipids, which can result in a reduction in the pH gradient between the intra- and extracellular space. Currently known mechanisms to reduce the toxicity of alkanes and alkenes include the use of ABC transporters from *Yarrowia lipolytica* expressed in *S. cerevisiae* for export (Chen et al. 2013). The heterologous expression of exporters resulted in 5 and 30 fold lower decane and undecane intracellular levels, respectively, which improved tolerance level to decane by 80-fold (Chen et al. 2013), further demonstrating the effectiveness of the use of efflux pumps. It has also been noted that deletion of the *acrAB* efflux pump genes in *E. coli* reduced tolerance to n-hexane and cyclohexane. Correspondingly, associated drug efflux pumps *marRAB*, when overexpressed, increased tolerance to n-hexane and cyclohexane (Aono et al. 1998). General stress response mechanisms such as heat shock protein expression has demonstrated some success in alleviating alkane tolerance; for example, expression of heterologous heat shock chaperone proteins in *E. coli* increased tolerance to n-hexane and octane (Nicolaou et al. 2010).

5.4.2.3 Organic Acids Toxicity

Organic acids are a diverse group of chemicals, with many being essential to life as building blocks or key intermediates in metabolism. However high concentrations of even generally innocuous chemicals can be toxic to the cell; therefore, while organic acids are generally non-toxic, accumulation can lead to cellular toxicity. One known mode of organic acids toxicity is attributed to the ability of undissociated acids to diffuse across the cell membrane freely as described in weak acids Sect. 5.2.2.2. This can both disrupt the internal cellular pH and the intracellular anion pool. To the contrary, however, some amino acids actually contribute toward increased tolerance to acidic conditions (Zhao and Bai 2009). An example of this is in the acid response systems 2 and 3 in *E. coli* as detailed above in Sect. 3.2.3. Similar to responses to other toxicities, when organic acids do inhibit cell function, *E coli* has also been known to respond to membrane disruption caused by organic acid stress by compensating for the perturbations via adjusting fatty acid and lipid content (Warnecke and Gill 2005). In order to increase tolerance to certain organic acids such as acetic acid and butyrate, some organisms are able to uptake and consume the acids through natural metabolism. For example, butyrate can be converted to butyryl-CoA and then neutral butanol, or acetic acid can be converted to acetyl-CoA for metabolism into numerous products [reviewed by (Nicolaou et al. 2010)]. Finally, some organic acids trigger chaperone protein expression in *S. cerevisiae*, likely to refold proteins destabilized by the presence of organic acid (Nicolaou et al. 2010).

5.4.2.4 Alcohols

Ethanol

Ethanol is currently a primary biofuel target, and combined with its historical role in fermented alcoholic beverages, it is not surprising there is extensive research on ethanol toxicity. Ethanol has been observed to induce damage by destabilizing cellular structures including proteins, membranes, and complexes important for transcription and translation. Ethanol intercalates into the membrane and increases membrane fluidity, which leads to interference of important cellular functions, particularly proton gradient and the overall electrostatic potential. These two gradients are harnessed by the cell to generate energy through respiration and utilized to perform active transport. To rectify this disruption of the cellular envelope and membrane, several modifications to cell wall and cell membrane construction are known to increase tolerance. For example in *E. coli*, augmentation with the overexpression of two fatty acid dehydratase genes, *fabA* and *des* from *Bacillus subtilis*, encoding for enzymes involved in de-saturating membrane fatty acids, resulted in an increase in ethanol tolerance (Jin et al. 2014). In *E. coli*, another detected membrane modification that increases tolerance toward ethanol is changes to the ratio of different phospholipid head groups (Nicolaou et al. 2010). The change from phosphatidyl-ethanolamine to diphosphatidyl-glycerol and phosphatidyl-glycerol potentially alters the membrane surface charge, which may reduce the membrane stress induced by ethanol (Nicolaou et al. 2010); the same work also reviewed how *S. cerevisiae* is likewise observed to increase the quantity of unsaturated fatty acids and ergosterols in the membrane in response to ethanol, both of which are predicted to have a stabilizing effect on membrane fluidity (Nicolaou et al. 2010). Membrane stability is also increased in response to ethanol via modifications in cell wall biosynthesis. In *E. coli*, overexpression of the *mur* operon genes, which are important for cell wall biosynthesis, has been shown to increase tolerance to ethanol by 160 % (Goodarzi et al. 2010). The cellular damage caused by ethanol, such as through oxidation and destabilization of protein conformations, can be alleviated through expression of heat shock chaperone proteins. Deletion of various HSPs has been shown to increase susceptibility to ethanol, whereas increased expression of *hsp70*, *hsp30* and *hsp12* in *S. cerevisiae* increase ethanol tolerance [as reviewed in (Nicolaou et al. 2010)]; overexpression of heat shock proteins *hsp1* and *hsp26* have also been shown to enhance ethanol tolerance in yeast (Zhao and Bai 2009). Furthermore, ethanol also interferes with transcription and translational machinery; it negatively impacts translation by inducing errors through misreading, stalling ribosomes, reduction in the effectiveness of rho dependent termination in *E. coli* (Haft et al. 2014), and interference with transcriptional termination and transcript elongation. Various beneficial mutations in the proteins involved in these processes have been identified and found to alleviate the negative impacts on transcription and translation imposed by ethanol; thereby, indicating this mode of toxicity may be readily avoidable with minor protein modifications (Haft et al. 2014).

Butanols

Butanol is highly toxic to microbial systems and can inhibit cellular function at very low concentrations. The modes of toxicity of butanol include interruption of membrane function and form, leading to leakiness of protons and other ions, which lowers the cellular proton motive force. High n-butanol concentrations also inhibit active transport of important nutrients such as glucose into the cell. Similar to ethanol stress, HSPs that act as folding chaperones have been shown to alleviate butanol stress by assisting in the repair of misfolded proteins. Examples include the *groESL* proteins in *Clostridium acetobutylicum*, increased expression of which increased tolerance to n-butanol by 85 % (Tomas et al. 2003). Increased expression of homologs of *groESL* and other known stress response proteins in other organisms have also been shown to increase tolerance to a variety of stressors [as reviewed by (Dunlop 2011)]. Modification of membrane fatty acid saturation levels has been identified as a potential mechanism for n-butanol tolerance (Nicolaou et al. 2010). This is parallel to the stress response seen in ethanol. For example, *C. acetobutylicum* ATCC824 was observed to increase the quantity of saturated fatty acids relative to unsaturated fatty acids in the membrane when exposed to n-butanol during exponential growth and when synthesizing butanol in stationary phase, which presumably stabilizes the membrane in the presence of butanol (Ezeji et al. 2010). In addition, the ratio of longer fatty acids also increased when *E. coli* was grown in the presence of butanol (Ezeji et al. 2010). One of the most ubiquitous means of detoxification in microbial systems is through efflux pumps. However, few efflux pumps have been shown to effectively increase tolerance to short-chain alcohols as discussed in (Dunlop 2011). One example is the overexpression of the *focA* gene, which encodes an efflux pump of formate in *E. coli*, which increased n-butanol tolerance by exporting the toxic n-butanol molecules from the cell (Reyes et al. 2012). AcrB is another well-characterized efflux pump of various substrates in *E. coli* and does not normally export n-butanol; protein engineering efforts successfully modified this pump to export n-butanol and increased tolerance in *E. coli* by 25 % when expressed (Jin et al. 2014).

Long-Chain Alcohols

Long-chain alcohols have not yet been produced in large quantities and the pathways of production are relatively diverse. Consequently, the modes of toxicity and mechanisms of tolerance are not well characterized. One study examined the influence of the chain length of alcohols on the cellular response of *E. coli*, and found that short-chain alcohols such as ethanol and butanol resulted in increased concentration of unsaturated fatty acids in the membrane while longer chain alcohols elicited the opposite effect (Weber and de Bont 1996). The divergent responses indicate that the modes of stabilizing the membrane differ as alcohol chain length changes, and the tolerance response also differs between organisms. At longer chain lengths the membrane fluidity is reduced, while when exposed to shorter length alcohols the membrane is fluidized to a greater degree (Weber and de Bont 1996).

5.5 Summary/Future Directions

The majority of our current knowledge on microbial response and adaptation to inhibitory conditions is derived from studies using laboratory strains, and may not directly translate to industrial strains. With the advent of omics technologies, one can characterize the cellular responses of industrial strains in relevant inhibitory conditions at the molecular level to gain insights into potential targets for rational strain engineering. In addition, inverse engineering approaches can be readily applied to industrial strains. For example, a producer of interest can be evolved at high temperatures relevant for simultaneous saccharification and fermentation to select for thermotolerant mutants. Since productivity and robustness are generally decoupled, and evolved mutants often exhibit reduced productivity, it is desirable to either rationally engineer the robust mutants to optimize productivity or to identify the mechanisms of tolerance to rationally engineer a good producer for better robustness. In the latter case, whole genome sequencing, transcriptomics, proteomics, and metabolomics tools can be applied to the evolved mutants to identify the molecular mechanisms responsible for the desired phenotype. With rapid advances in new and broad-host-range genomic editing tools, the knowledge gained can be applied to engineer more robust industrial producer.

References

Allen SA, Clark W, McCaffery JM, Cai Z, Lanctot A, Slininger PJ, Liu ZL, Gorsich SW (2010) Furfural induces reactive oxygen species accumulation and cellular damage in Saccharomyces cerevisiae. Biotechnol Biofuels 3

Almario MP, Reyes LH, Kao KC (2013) Evolutionary engineering of Saccharomyces cerevisiae for enhanced tolerance to hydrolysates of lignocellulosic biomass. Biotechnol Bioeng 110 (10):2616–2623. doi:10.1002/bit.24938

Almeida JRM, Modig T, Petersson A, Hahn-Hagerdal B, Liden G, Gorwa-Grauslund MF (2007) Increased tolerance and conversion of inhibitors in lignocellulosic hydrolysates by Saccharomyces cerevisiae. J Chem Technol Biotechnol 82(4):340–349. doi:10.1002/Jctb.1676

Almeida JR, Roder A, Modig T, Laadan B, Liden G, Gorwa-Grauslund MF (2008) NADH- vs NADPH-coupled reduction of 5-hydroxymethyl furfural (HMF) and its implications on product distribution in Saccharomyces cerevisiae. Appl Microbiol Biotechnol 78(6):939–945. doi:10.1007/s00253-008-1364-y

Alper H, Moxley J, Nevoigt E, Fink GR, Stephanopoulos G (2006) Engineering yeast transcription machinery for improved ethanol tolerance and production. Science 314(5805):1565–1568

Altendorf K, Booth I, Gralla J, Greie J, Rosenthal A, Wood J (2009) Osmotic Stress. EcoSal Plus. doi:10.1128/ecosalplus.5.4.5

An MZ, Tang YQ, Mitsumasu K, Liu ZS, Shigeru M, Kenji K (2011) Enhanced thermotolerance for ethanol fermentation of Saccharomyces cerevisiae strain by overexpression of the gene coding for trehalose-6-phosphate synthase. Biotechnol Lett 33(7):1367–1374. doi:10.1007/s10529-011-0576-x

Aono R, Tsukagoshi N, Yamamoto M (1998) Involvement of outer membrane protein TolC, a possible member of the mar-sox regulon, in maintenance and improvement of organic solvent tolerance of Escherichia coli K-12. J Bacteriol 180(4):938–944

Ask M, Bettiga M, Mapelli V, Olsson L (2013a) The influence of HMF and furfural on redox-balance and energy-state of xylose-utilizing Saccharomyces cerevisiae. Biotechnol Biofuels 6(1):22. doi:10.1186/1754-6834-6-22

Ask M, Mapelli V, Hock H, Olsson L, Bettiga M (2013b) Engineering glutathione biosynthesis of Saccharomyces cerevisiae increases robustness to inhibitors in pretreated lignocellulosic materials. Microb Cell Fact 12:87. doi:10.1186/1475-2859-12-87

Attfield PV (1997) Stress tolerance: the key to effective strains of industrial baker's yeast. Nat Biotechnol 15(13):1351–1357. doi:10.1038/Nbt1297-1351

Basak S, Geng H, Jiang R (2014) Rewiring global regulator cAMP receptor protein (CRP) to improve E. coli tolerance towards low pH. J Biotechnol 173:68–75. doi:10.1016/j.jbiotec.2014.01.015

Bauer BE, Rossington D, Mollapour M, Mamnun Y, Kuchler K, Piper PW (2003) Weak organic acid stress inhibits aromatic amino acid uptake by yeast, causing a strong influence of amino acid auxotrophies on the phenotypes of membrane transporter mutants. Eur J Biochem 270 (15):3189–3195. doi:10.1046/j.1432-1033.2003.03701.x

Beales N (2004) Adaptation of microorganisms to cold temperatures, weak acid preservatives, low pH, and osmotic stress: a review. Compr Rev Food Sci Food Safety 3(1):1–20. doi:10.1111/j.1541-4337.2004.tb00057.x

Boot IR, Cash P, O'Byrne C (2002) Sensing and adapting to acid stress. Antonie Van Leeuwenhoek 81(1–4):33–42

Booth IR (1989) The preservation of foods by low pH in Mechanisms of action of food preservation procedures

Castanie-Cornet MP, Penfound TA, Smith D, Elliott JF, Foster JW (1999) Control of acid resistance in Escherichia coli. J Bacteriol 181(11):3525–3535

Cayley S, Record MT Jr (2004) Large changes in cytoplasmic biopolymer concentration with osmolality indicate that macromolecular crowding may regulate protein-DNA interactions and growth rate in osmotically stressed Escherichia coli K-12. J Mol Recognit 17(5):488–496. doi:10.1002/jmr.695

Chang YY, Cronan JE (1999) Membrane cyclopropane fatty acid content is a major factor in acid resistance of Escherichia coli. Mol Microbiol 33(2):249–259. doi:10.1046/j.1365-2958.1999.01456.x

Chen RE, Thorner J (2007) Function and regulation in MAPK signaling pathways: lessons learned from the yeast Saccharomyces cerevisiae. Biochim Biophys Acta 1773(8):1311–1340

Chen B, Ling H, Chang MW (2013) Transporter engineering for improved tolerance against alkane biofuels in Saccharomyces cerevisiae. Biotechnol Biofuels 6:21. doi:10.1186/1754-6834-6-21

Cheroutre-Vialette M, Lebert I, Hebraud M, Labadie JC, Lebert A (1998) Effects of pH or a(w) stress on growth of Listeria monocytogenes. Int J Food Microbiol 42(1–2):71–77 S0168-1605(98)00064-6 [pii]

Choi SH, Baumler DJ, Kaspar CW (2000) Contribution of dps to acid stress tolerance and oxidative stress tolerance in Escherichia coli O157:H7. Appl Environ Microbiol 66(9):3911–3916

Demeke MM, Dietz H, Li Y, Foulquie-Moreno MR, Mutturi S, Deprez S, Den Abt T, Bonini BM, Liden G, Dumortier F, Verplaetse A, Boles E, Thevelein JM (2013) Development of a D-xylose fermenting and inhibitor tolerant industrial Saccharomyces cerevisiae strain with high performance in lignocellulose hydrolysates using metabolic and evolutionary engineering. Biotechnol Biofuels 6(1):89. doi:10.1186/1754-6834-6-89

Dunlop MJ (2011) Engineering microbes for tolerance to next-generation biofuels. Biotechnol Biofuels 4(1):32. doi:10.1186/1754-6834-4-32

Ezeji T, Milne C, Price ND, Blaschek HP (2010) Achievements and perspectives to overcome the poor solvent resistance in acetone and butanol-producing microorganisms. Appl Microbiol Biotechnol 85(6):1697–1712. doi:10.1007/s00253-009-2390-0

Gaida SM, Al-Hinai MA, Indurthi DC, Nicolaou SA, Papoutsakis ET (2013) Synthetic tolerance: three noncoding small RNAs, DsrA, ArcZ and RprA, acting supra-additively against acid stress. Nucleic Acids Res 41(18):8726–8737. doi:10.1093/nar/gkt651

Gershenzon J, Dudareva N (2007) The function of terpene natural products in the natural world. Nat Chem Biol 3(7):408–414. doi:10.1038/nchembio.2007.5

Gibson BR, Lawrence SJ, Leclaire JPR, Powell CD, Smart KA (2007) Yeast responses to stresses associated with industrial brewery handling. FEMS Microbiol Rev 31(5):535–569. doi:10.1111/j.1574-6976.2007.00076.x

Goodarzi H, Bennett BD, Amini S, Reaves ML, Hottes AK, Rabinowitz JD, Tavazoie S (2010) Regulatory and metabolic rewiring during laboratory evolution of ethanol tolerance in E. coli. Mol Syst Biol 6:378. doi:10.1038/msb.2010.33

Gorsich SW, Dien BS, Nichols NN, Slininger PJ, Liu ZL, Skory CD (2006) Tolerance to furfural-induced stress is associated with pentose phosphate pathway genes ZWF1, GND1, RPE1, and TKL1 in Saccharomyces cerevisiae. Appl Microbiol Biotechnol 71(3):339–349. doi:10.1007/s00253-005-0142-3

Haft RJ, Keating DH, Schwaegler T, Schwalbach MS, Vinokur J, Tremaine M, Peters JM, Kotlajich MV, Pohlmann EL, Ong IM, Grass JA, Kiley PJ, Landick R (2014) Correcting direct effects of ethanol on translation and transcription machinery confers ethanol tolerance in bacteria. Proc Natl Acad Sci USA 111(25):E2576–E2585. doi:10.1073/pnas.1401853111

Hohmann S (2002) Osmotic stress signaling and osmoadaptation in yeasts. Microbiol Mol Biol Rev 66(2):300–372

Hottiger T, De Virgilio C, Hall MN, Boller T, Wiemken A (1994) The role of trehalose synthesis for the acquisition of thermotolerance in yeast. II. Physiological concentrations of trehalose increase the thermal stability of proteins in vitro. Eur J Biochem 219(1–2):187–193

Jin H, Chen L, Wang J, Zhang W (2014) Engineering biofuel tolerance in non-native producing microorganisms. Biotechnol Adv 32(2):541–548. doi:10.1016/j.biotechadv.2014.02.001

Jonsson LJ, Alriksson B, Nilvebrant NO (2013) Bioconversion of lignocellulose: inhibitors and detoxification. Biotechnol Biofuels 6(1):16. doi:10.1186/1754-6834-6-16

Jordan KN, Oxford L, O'Byrne CP (1999) Survival of low-pH stress by Escherichia coli O157:H7: correlation between alterations in the cell envelope and increased acid tolerance. Appl Environ Microbiol 65(7):3048–3055

Kapteyn JC, ter Riet B, Vink E, Blad S, De Nobel H, Van Den Ende H, Klis FM (2001) Low external pH induces HOG1-dependent changes in the organization of the Saccharomyces cerevisiae cell wall. Mol Microbiol 39(2):469–479 mmi2242 [pii]

Kaushik JK, Bhat R (2003) Why is trehalose an exceptional protein stabilizer? An analysis of the thermal stability of proteins in the presence of the compatible osmolyte trehalose. J Biol Chem 278(29):26458–26465. doi:10.1074/jbc.M300815200

Khankal R, Chin JW, Ghosh D, Cirino PC (2009) Transcriptional effects of CRP* expression in Escherichia coli. J Biol Eng 3:13. doi:10.1186/1754-1611-3-13

Kim D, Hahn JS (2013) Roles of the Yap1 transcription factor and antioxidants in Saccharomyces cerevisiae's tolerance to furfural and 5-hydroxymethylfurfural, which function as thiol-reactive electrophiles generating oxidative stress. Appl Environ Microbiol 79(16):5069–5077. doi:10.1128/Aem.00643-13

Klein-Marcuschamer D, Stephanopoulos G (2008) Assessing the potential of mutational strategies to elicit new phenotypes in industrial strains. Proc Natl Acad Sci USA 105(7):2319–2324. doi:10.1073/pnas.0712177105

Klinke HB, Thomsen AB, Ahring BK (2004) Inhibition of ethanol-producing yeast and bacteria by degradation products produced during pre-treatment of biomass. Appl Microbiol Biotechnol 66 (1):10–26. doi:10.1007/s00253-004-1642-2

Koppram R, Albers E, Olsson L (2012) Evolutionary engineering strategies to enhance tolerance of xylose utilizing recombinant yeast to inhibitors derived from spruce biomass. Biotechnol Biofuels 5(1):32. doi:10.1186/1754-6834-5-32

Kramer R (2010) Bacterial stimulus perception and signal transduction: response to osmotic stress. Chem Rec 10(4):217–229. doi:10.1002/tcr.201000005

Laadan B, Almeida JR, Radstrom P, Hahn-Hagerdal B, Gorwa-Grauslund M (2008) Identification of an NADH-dependent 5-hydroxymethylfurfural-reducing alcohol dehydrogenase in Saccharomyces cerevisiae. Yeast 25(3):191–198. doi:10.1002/yea.1578

Larsson S, Palmqvist E, Hahn-Hagerdal B, Tengborg C, Stenberg K, Zacchi G, Nilvebrant NO (1999) The generation of fermentation inhibitors during dilute acid hydrolysis of softwood. Enzyme and Microbial Technology 24(3–4):151–159. doi:10.1016/S0141-0229(98)00101-X

Lin J, Lee IS, Frey J, Slonczewski JL, Foster JW (1995) Comparative analysis of extreme acid survival in Salmonella typhimurium, Shigella flexneri, and Escherichia coli. J Bacteriol 177 (14):4097–4104

Lindquist S (1992) Heat-shock proteins and stress tolerance in microorganisms. Curr Opin Genet Dev 2(5):748–755

Ling H, Teo W, Chen B, Leong SS, Chang MW (2014) Microbial tolerance engineering toward biochemical production: from lignocellulose to products. Curr Opin Biotechnol 29:99–106

Mager WH, De Kruijff AJ (1995) Stress-induced transcriptional activation. Microbiol Rev 59 (3):506–531

Martinez-Pastor MT, Marchler G, Schuller C, Marchler-Bauer A, Ruis H, Estruch F (1996) The Saccharomyces cerevisiae zinc finger proteins Msn2p and Msn4p are required for transcriptional induction through the stress response element (STRE). EMBO J 15(9):2227–2235

Mendoza I, Quintero FJ, Bressan RA, Hasegawa PM, Pardo JM (1996) Activated calcineurin confers high tolerance to ion stress and alters the budding pattern and cell morphology of yeast cells. J Biol Chem 271(38):23061–23067

Meury J (1988) Glycine betaine reverses the effects of osmotic-stress on DNA Replication and cellular division in Escherichia coli. Arch Microbiol 149(3):232–239. doi:10.1007/Bf00422010

Miller EN, Jarboe LR, Yomano LP, York SW, Shanmugam KT, Ingram LO (2009) Silencing of NADPH-dependent oxidoreductase genes (yqhD and dkgA) in furfural-resistant ethanologenic Escherichia coli. Appl Environ Microbiol 75(13):4315–4323. doi:10.1128/AEM.00567-09

Morano KA, Liu PC, Thiele DJ (1998) Protein chaperones and the heat shock response in Saccharomyces cerevisiae. Curr Opin Microbiol 1(2):197–203 S1369-5274(98)80011-8 [pii]

Morano KA, Grant CM, Moye-Rowley WS (2012) The response to heat shock and oxidative stress in Saccharomyces cerevisiae. Genetics 190(4):1157–1195 10.1534/genetics.111.128033

Murdock L, Burke T, Coumoundouros C, Culham DE, Deutch CE, Ellinger J, Kerr CH, Plater SM, To E, Wright G, Wood JM (2014) Analysis of strains lacking known osmolyte accumulation mechanisms reveals contributions of osmolytes and transporters to protection against abiotic stress. Appl Environ Microbiol 80(17):5366–5378. doi:10.1128/AEM.01138-14

Nicolaou SA, Gaida SM, Papoutsakis ET (2010) A comparative view of metabolite and substrate stress and tolerance in microbial bioprocessing: from biofuels and chemicals, to biocatalysis and bioremediation. Metab Eng 12(4):307–331. doi:10.1016/j.ymben.2010.03.004

Parawira W, Tekere M (2011) Biotechnological strategies to overcome inhibitors in lignocellulose hydrolysates for ethanol production: review. Crit Rev Biotechnol 31(1):20–31. doi:10.3109/07388551003757816

Park YK, Bearson B, Bang SH, Bang IS, Foster JW (1996) Internal pH crisis, lysine decarboxylase and the acid tolerance response of Salmonella typhimurium. Mol Microbiol 20(3):605–611. doi:10.1046/j.1365-2958.1996.5441070.x

Parrou JL, Teste MA, Francois J (1997) Effects of various types of stress on the metabolism of reserve carbohydrates in Saccharomyces cerevisiae: genetic evidence for a stress-induced recycling of glycogen and trehalose. Microbiology-Uk 143:1891–1900

Peralta-Yahya PP, Zhang F, del Cardayré SB, Keasling JD (2012) Microbial engineering for the production of advanced biofuels. Nature 488(7411):320–328. doi:10.1038/nature11478

Petersson A, Almeida JR, Modig T, Karhumaa K, Hahn-Hagerdal B, Gorwa-Grauslund MF, Liden G (2006) A 5-hydroxymethyl furfural reducing enzyme encoded by the Saccharomyces cerevisiae ADH6 gene conveys HMF tolerance. Yeast 23(6):455–464. doi:10.1002/yea.1370

Portillo MD, Saadeddin A (2014) Recent trends in ionic liquid (IL) tolerant enzymes and microorganisms for biomass conversion. Crit Rev Biotechnol. doi:10.3109/07388551.2013.843069

Quijano G, Couvert A, Amrane A (2010) Ionic liquids: applications and future trends in bioreactor technology. Bioresour Technol 101(23):8923–8930. doi:10.1016/j.biortech.2010.06.161

Raja N, Goodson M, Smith DG, Rowbury RJ (1991) Decreased DNA damage by acid and increased repair of acid-damaged DNA in acid-habituated Escherichia coli. J Appl Bacteriol 70(6):507–511

Reyes LH, Almario MP, Winkler J, Orozco MM, Kao KC (2012) Visualizing evolution in real time to determine the molecular mechanisms of n-butanol tolerance in Escherichia coli. Metab Eng 14(5):579–590. doi:10.1016/j.ymben.2012.05.002

Richard H, Foster JW (2004) Escherichia coli glutamate- and arginine-dependent acid resistance systems increase internal pH and reverse transmembrane potential. J Bacteriol 186(18):6032–6041. doi:10.1128/JB.186.18.6032-6041.2004

Roth WG, Leckie MP, Dietzler DN (1985) Osmotic stress drastically inhibits active transport of carbohydrates by Escherichia coli. Biochem Biophys Res Commun 126(1):434–441 0006-291X(85)90624-2 [pii]

Russell JB (1992) Another explanation for the toxicity of fermentation acids at low pH: anion accumulation versus uncoupling. J Appl Bacteriol 73(5):363–370. doi:10.1111/j.1365-2672.1992.tb04990.x

Saha BC (2003) Hemicellulose bioconversion. J Ind Microbiol Biotechnol 30(5):279–291. doi:10.1007/s10295-003-0049-x

Saito H, Posas F (2012) Response to hyperosmotic stress. Genetics 192(2):289–318. doi:10.1534/genetics.112.140863

Shahsavarani H, Sugiyama M, Kaneko Y, Chuenchit B, Harashima S (2012) Superior thermotolerance of Saccharomyces cerevisiae for efficient bioethanol fermentation can be achieved by overexpression of RSP5 ubiquitin ligase. Biotechnol Adv 30(6):1289–1300. doi:10.1016/j.biotechadv.2011.09.002

Shi DJ, Wang CL, Wang KM (2009) Genome shuffling to improve thermotolerance, ethanol tolerance and ethanol productivity of Saccharomyces cerevisiae. J Ind Microbiol Biotechnol 36 (1):139–147. doi:10.1007/s10295-008-0481-z

Sikkema J, De Bont JA, Poolman B (1995) Mechanisms of membrane toxicity of hydrocarbons. Microbiol Rev 59(2):201–222

Steen EJ, Kang Y, Bokinsky G, Hu Z, Schirmer A, McClure A, del Cardayré SB, Keasling JD (2010) Microbial production of fatty-acid-derived fuels and chemicals from plant biomass. Nature 463(7280):559–562. doi:10.1038/nature08721

Terada H (1990) Uncouplers of oxidative phosphorylation. Environ Health Perspect 87:213–218

Thevelein JM (1994) Signal transduction in yeast. Yeast 10(13):1753–1790

Tomas CA, Welker NE, Papoutsakis ET (2003) Overexpression of groESL in Clostridium acetobutylicum results in increased solvent production and tolerance, prolonged metabolism, and changes in the cell's transcriptional program. Appl Environ Microbiol 69(8):4951–4965

Unnikrishnan I, Miller S, Meinke M, LaPorte DC (2003) Multiple positive and negative elements involved in the regulation of expression of GSY1 in Saccharomyces cerevisiae. J Biol Chem 278(29):26450–26457. doi:10.1074/jbc.M211808200

Vancov T, Alston AS, Brown T, McIntosh S (2012) Use of ionic liquids in converting lignocellulosic material to biofuels. Renewable Energy 45:1–6. doi:10.1016/j.renene.2012.02.033

Vogel J, Wagner EG (2007) Target identification of small noncoding RNAs in bacteria. Curr Opin Microbiol 10(3):262–270

Warnecke T, Gill RT (2005) Organic acid toxicity, tolerance, and production in Escherichia coli biorefining applications. Microb Cell Fact 4(1):25. doi:10.1186/1475-2859-4-25

Weber FJ, de Bont JA (1996) Adaptation mechanisms of microorganisms to the toxic effects of organic solvents on membranes. Biochimica et Biophysica Acta (BBA)-Reviews on Biomembranes 1286 (3):225–245

Winkler JD, Garcia C, Olson M, Callaway E, Kao KC (2014) Evolved osmotolerant Escherichia coli mutants frequently exhibit defective N-acetylglucosamine catabolism and point mutations in cell shape-regulating protein MreB. Appl Environ Microbiol 80(12):3729–3740. doi:10.1128/AEM.00499-14

Wood JM (1999) Osmosensing by bacteria: signals and membrane-based sensors. Microbiol Mol Biol Rev 63(1):230–262

Wood JM (2015) Bacterial responses to osmotic challenges. J Gen Physiol 145(5):381–388. doi:10.1085/jgp.201411296

Yamamoto A, Mizukami Y, Sakurai H (2005) Identification of a novel class of target genes and a novel type of binding sequence of heat shock transcription factor in Saccharomyces cerevisiae. J Biol Chem 280(12):11911–11919

Yuk HG, Marshall DL (2004) Adaptation of Escherichia coli O157: H7 to pH alters membrane lipid composition, verotoxin secretion, and resistance to simulated gastric fluid acid. Appl Environ Microbiol 70(6):3500–3505. doi:10.1128/Aem.70.6.3500-3505.2004

Zhang H, Chong H, Ching CB, Jiang R (2012) Random mutagenesis of global transcription factor cAMP receptor protein for improved osmotolerance. Biotechnol Bioeng 109(5):1165–1172. doi:10.1002/bit.24411

Zhao XQ, Bai FW (2009) Mechanisms of yeast stress tolerance and its manipulation for efficient fuel ethanol production. J Biotechnol 144(1):23–30. doi:10.1016/j.jbiotec.2009.05.001

Chapter 6
Biobased Technology Commercialization: The Importance of Lab to Pilot Scale-up

James P. Wynn, Robert Hanchar, Susanne Kleff, David Senyk and Tonya Tiedje

Abstract The transition of a renewable biobased technology from lab scale (at ≤10 L) to pilot scale (≥1000 L) is a critical, though costly and intimidating, step in technology derisking. Once a process has been demonstrated at pilot scale (≥1000 L), there should be a high degree of confidence that the process can be performed successfully at commercial scale. Understanding (i) the value of pilot-scale operation; (ii) process features that can be addressed prior to pilot plant scale-up; (iii) those that have to be addressed at pilot scale; and (iv) the best approaches to pilot scale-up can make this aspect of technology derisking less intimidating, less time consuming, less expensive, and ultimately an easier sell to either management or investors.

6.1 Technology Derisking

Any new technology developed at laboratory scale has a high degree of risk associated with it. Initially, these risks are centered on the ability of the technology to meet key performance milestones that indicate whether or not it has the capability of being economically successful. As potential risks are eliminated from a nascent technology, the chance of its success takes shape and its value increases. Derisking at laboratory scale is only one facet of technology development, however, and in order to become commercially viable, any biobased technology must also demonstrate that it is robust and scalable to commercially relevant size. The first step in the demonstration of process robustness and scalability is the transition from laboratory to pilot-scale operation.

J.P. Wynn (✉) · R. Hanchar · S. Kleff · D. Senyk · T. Tiedje
MBI, 3815 Technology Blvd., Lansing, MI 48910, USA
e-mail: jimwynn@abmauri.com
URL: http://www.mbi.org

S. Van Dien (ed.), *Metabolic Engineering for Bioprocess Commercialization*,
DOI 10.1007/978-3-319-41966-4_6

6.2 What Is Pilot Scale?

All fermentation biologists and biotechnologists have a general understanding that "pilot scale" is an intermediate step between lab-scale process development and full-scale commercial production. If pressed, however, the exact definition of what constitutes pilot-scale fermentation can widely differ. In some cases, scientists consider 10–100 L fermentation as pilot scale, whereas for others (including the authors) pilot scale is considerably larger and would constitute fermentation vessels with a working volume ≥1000 L. For the purposes of this chapter, the authors will define pilot-scale fermentation at ≥1000 L, because at this scale many of the issues that can plague or even derail a biobased technology as it transitions from the R&D lab (typically <1 L to ≈10 L working volume) to commercial scale >10,000 L can be tested and addressed.

6.3 Why Transition to Pilot Scale?

At MBI, many of the processes that are scaled up to pilot scale (i.e., >1000 L working volume) are biobased fuels and chemicals (rather than therapeutic molecules) that are intended for very large, commercial scale (≥100 m^3) production. Therefore, although transition to a pilot scale may be a significant increase in operating volume, even this may be an intermediate stage with another intermediate scale-up to demo-scale (≥50,000 L) required before final commercial production scale is achieved. Even if a demo-scale trial is necessary, the transition from lab to pilot scale remains a critical step—one that removes significant risk and increases the credibility (and therefore value) of the biobased technology. A successful demonstration at pilot scale gives a greater confidence that the developed process is fundamentally scalable before the capital expense involved in the building of a demonstration or commercial-scale facility is invested.

The lab to pilot-scale transition is not a trivial undertaking in terms of effort or cost. Consequently, early in a project this process should not be rushed into without due consideration. Scale-up of a fermentation technology too early in its development cycle can incur significant cost and yet offer few benefits other than confirming that the technology is at too early a state of development to warrant scale-up efforts. There are some situations where scale-up of an early stage fermentation technology can be appropriate, such as to confirm that a biobased compound has the required functionality; however, even in those situations the risks associated are high and should be recognized. The current chapter aims to guide the reader in deciding when transitioning a process from laboratory to pilot scale is appropriate and what advantages can be gained due to the scale-up activity.

Despite the challenges and expense involved, the transition from lab to pilot-scale is an essential stage in the commercialization of fermentation-based technology (Junker 2004), resulting in a far greater degree of confidence—within the technical R&D team, management, potential customers, and investors—that the process is robust, scalable, and commercially viable.

6.4 Addressing Process Development Issues Before Technology Scale-up

There are many challenges that need to be addressed when a biobased process, developed at lab scale, needs to be transitioned to commercial scale production. While many of these challenges can only be realistically addressed at pilot scale, others can be addressed even before scale-up efforts are initiated. The authors recommend addressing those issues that can be tackled at lab scale before embarking on pilot-scale testing. Experiments at 1–10 L (lab scale) are much easier and far less expensive than at ≥1000 L scale.

Factors that can be addressed before scale-up is attempted include cost of fermentation medium and seed train development. Although these can be addressed at lab scale, they are frequently overlooked at the early stage because they are not critical factors that impact the ability to operate and carry out laboratory strain and process development. Often, management and investors are less interested in these aspects of process development than they are in fermentation performance, since they are less tangible than metrics like productivity, titer, and yield (the exception being, perhaps, those management or investor groups who have been blindsided by these issues in the past). Addressing these factors earlier in the process development cycle is valuable, as it can avoid costly delays during subsequent scale-up.

6.4.1 Cost of Medium Ingredients

During the early stages of biobased technology development, milestones and targets are frequently focused on fermentation performance parameters: product titer, yield, and productivity. This makes perfect sense, as continued development of a process that cannot meet commercially viable performance targets would be foolhardy. However, in the initial race to meet fermentation targets, the cost of the medium components is often overlooked. At lab scale, the cost of the medium is irrelevant, when comparing other costs (scientists' time, etc), and tends to be neglected. When producing biobased chemicals (especially fuels), the "cost of manufacture" targets are very low compared with food or drug products; however, the potential sales volumes are far higher, making them economically interesting targets. As a result of the low selling price of these chemicals (usually in the $0.4–$2.0/lb. range), the medium cost needs to be monitored closely. Many academic laboratories routinely use a rich, complex medium, containing yeast extract or peptone as a complex nitrogen and nutrient source. These complex media are (i) simple to prepare, (ii) ensure good growth of most microorganisms (simplifying early experimentation), and (iii) facilitate reaching initial process performance targets. Therefore, these complex media are appropriate for the initial testing of production organisms and culture conditions in a time efficient manner. When technologies are transferred from university laboratories to biotechnology companies, these artefacts of the initial proof of concept research work are often maintained. Therefore, even as other elements of the technology are developed (culture conditions, feed profiles, etc) the use of complex N-sources continues. However, both yeast extract and peptone are expensive medium components (in the range of \$2–9/lb) compared to dextrose (\$0.10–0.25/lb, depending on the price of corn). If yeast extract or peptone are used in culture media as a major nutrient (as the sole or major nitrogen source), then the fermentation medium will not be economically viable. Low levels of yeast extract are tolerated in some industrial media (<1 g/L) as a source of vitamins and less well-defined "growth factors" to maximize growth of more fastidious production hosts, but the levels should always be minimized. It is remarkable how often companies initiate the scale-up of their fermentation—having invested significant research dollars and time in process development—only to be shocked by the cost when ≥1000 L of medium are needed for pilot testing. This can require more time and money to re-optimize the fermentation medium/process to remove the yeast extract (or other cost prohibitive medium ingredient) before scale-up can proceed.

Yet another complication of using high-cost, complex medium components like yeast extract, or even low-cost, complex medium components like corn steep liquor (CSL), is that a greater burden is placed on the product separation and purification portion of the developing technology (see Sect. 5.4). Simple defined media, lacking complex, and poorly defined nutrient sources are ideal for commercial production. These media are generally inexpensive and simplify product purification. Although developing a defined medium for a particular production host may or may not be possible (depending on the host organism being employed), in all cases efforts to use the minimum amount of complex medium components should be a focus of process development and should be addressed before process scale-up is considered.

It is becoming more common, in this "post-genomic" world, to use metabolic engineering and synthetic biology to increase target chemical production by microbial cells. Sometimes these efforts are to increase production of a chemical produced naturally by the microorganism, and at other time these efforts are to produce chemicals in "preferred" production hosts that do not naturally produce the chemical. Indeed, it is rare that unmodified production hosts are used in the manufacture of a non-food, biobased chemical. Again, in striving to develop and optimize pathways to achieve fermentation production performance, small but critical flaws in a process can be perpetuated, rendering the technologies non-scalable. The easiest of these to avoid—and can be resolved long before scale-up—is the use of expensive chemical inducers to trigger target chemical production. Although, in some cases, constitutive strong promoters can be used (resulting in growth-associated production of the target chemical), this is often not the preferred route. Many of the target chemicals, particularly compounds not naturally synthesized by the production host, are toxic to the host and decrease cell growth and product formation, especially at the high productivities and titers required for biobased chemical or fuel production. This toxicity can be mitigated by the use of an inducible promoter that triggers product formation after induction of the culture. In this manner, cell growth and product formation can be separated; the target chemical can only be produced once active cell growth is complete. Separation of cell growth and product formation is often effective in mitigating the toxicity that can be seen in actively growing cells. While many inducible promoters are available, one of the most convenient and commonly used (certainly in academic laboratories) is the lac promoter, induced using IPTG. Like yeast extract, the cost of addition of IPTG in lab scale cultures is insignificant and often goes unrecognized or is simply overlooked. When ≥1000 L of fermentation broth must be induced, the true cost of this inducer can suddenly come into sharp focus, especially with the prospect of further scale-up by two more orders of magnitude in terms of volume. The selection of (or even need for) an appropriate inducible promoter can be undertaken at any stage in the process development, but it is often pushed aside by the more pressing issues of fermentation performance (titer, productivity, and yield). Yet the nature of the promoter employed will need to be resolved before scale-up to pilot fermentors (and beyond) can be effectively and economically achieved.

Generally, medium ingredient cost and source should be considered at laboratory scale before embarking on scale-up efforts. As well as using complex medium components and expensive heterologous gene expression inducers, it is routine to use ACS grade (or higher) in a research laboratory setting. High purity, research grade chemicals are rarely cost effective at commercial scale and will need to be replaced by lower (industrial or food) grade chemicals that are lower in price and available in the quantities necessary for industrial-scale production. Confirmation that lower grade chemicals are compatible with the fermentation process (and the product separation and purification, *see below)* is required once scale-up is initiated. It is advisable to confirm the compatibility of industrial grade chemicals at lab scale before using them at pilot scale or above. It is a good rule of thumb that the biobased technology should be relatively well defined before scale-up is initiated, carrying out process development activities such as media, pH, or temperature optimization at pilot or production scale is ill-advised and should be avoided whenever possible.

6.4.2 Seed Train Development

Like medium cost, the seed train is another facet of fermentation often overlooked during the initial stages of technology development. The potential impact of this factor on large-scale production is not immediately obvious in the same degree as achieving cell density, product titer, and productivity targets. At laboratory scale, it is easy to generate sufficient inoculum volume (typically 2–10 % of production fermentor working volume) using a single shake flask culture (i.e., a 50 mL culture in a 250 mL shake flask can provide a 5 % v/v inoculum for a 10 L fermentor or a 10 % v/v inoculum for a fermentor ≤ 5 L). When the volume of the production tank is ≥1000 L, however, the generation of a 10 % (or even 5 %) seed inoculum is infeasible using a single shake flask step. Seed trains for industrial fermentations usually involve multiple stages, starting in a shake flask (or similar) and progressing through a series of progressively larger fermentors, each inoculated with 1–10 % seed volume from the preceding seed stage, to a final seed tank that is capable of supplying the necessary volume of seed culture to one (or more) production tanks.

Many permutations are possible, when developing a multistage seed train. Here are several factors to consider

i. *Medium composition*: Is this constant throughout the seed train or is the seed medium different than the production medium?
ii. *Inoculum size at each transfer.* Is a constant inoculum volume to be employed throughout the seed train, or could this vary? Use of a smaller inoculum volume (1 or 2 % v/v c.f. 10 % v/v) can result in longer incubation times. This may be permissible in a seed vessel whereas it may be less desirable in a production vessel. Seed tanks run in parallel to production batches and so do not necessarily determine the overall volumetric productivity of the fermentation facility.
iii. *Transfer criteria.* How is it decided when to inoculate a production tank or seed tank from the preceding seed stage? Transfer criteria can be as simple as a defined time (e.g., after 16 h) or may be related to phase of growth of the seed culture. If the latter, the timing of culture transfer can be based on any of a number of factors including; optical density (OD), glucose consumption, oxygen utilization rate (OUR) or a nutrient limitation. Generally, a physiologically relevant transfer criterion provides more consistent performance than a time-based transfer, particularly during scale-up. Although seed train development is a portion of the overall fermentation process that is rarely discussed in the peer-reviewed literature, there is anecdotal evidence that the physiological state of a culture at transfer can impact the performance of subsequent culture stages and the final production tank.
iv. *Strain stability.* If the production strain is a metabolically engineered microorganism and the heterologous genes are maintained on plasmids, then strain stability can be a significant issue. If the plasmid is maintained via selective pressure (e.g., based on antibiotic resistance), is the selection agent to be maintained throughout the seed train and production? If so, the impact on medium cost should be evaluated. If not, then the decision must be made on which stages of the seed train the selection agent will be employed. Understanding the strain stability of the production host becomes critical as the length and number of cultures stages (and therefore number of cell generations) in the seed train increase.

v. *Contamination mitigation.* In a single stage (relatively short) seed train contamination is not usually a concern; however, once multiple seed stages are implemented (up to five stages at different volumes are not unusual)—each stage involving a culture transfer step—the potential for culture contamination proportionately increases. A contamination rate of 1 in 20 (5 %) using a single seed may be acceptable, but if this contamination rate is perpetuated for each step in a multistage seed train, the contamination rate may be unacceptably high. A five-stage seed train with a 5 % contamination rate at each step results in a >20 % chance of contamination by the time the production vessel is reached. Aerobic fermentations, operating at mesophilic temperatures (20–40 °C) and at neutral pH in the presence of readily assimilated carbon and nitrogen sources, are the most prone to contamination. Anything that can shift the fermentation conditions away from this "ideal growth" condition will help prevent contamination and render the seed train (and the whole fermentation process) more robust.

None of the above factors requires testing at pilot scale. The seed train progression can be simulated in the R&D lab using shake flasks and stirred tank fermentors, using just the appropriate volume of each stage to inoculate the next, all the way to the lab-scale production vessel. Nevertheless, it is common for early stage biotechnology companies to be on the verge of scaling up before any attention is given to how the seed inoculum should be generated.

It is the authors' opinion that the issues described above should be addressed, or at least considered, during the early phases of process development and definition. Addressing these factors early in technology development may prevent a lab-scale fermentation that is either impossible to scale-up or that requires significant modification and adaptation in order to make it compatible with transitions to pilot and subsequently commercial fermentation volumes.

6.5 Factors to Be Addressed at Pilot Scale

Other fermentation parameters are not so easily defined or evaluated, and it is in the testing of these that fermentors ≥1000 L can be invaluable. These parameters are less predictably evaluated at volumes of 1–30 L in glass lab bench fermentors, which is why the authors prefer to define pilot scale as ≥1000 L.

Some fermentation groups highlight the fact that they do not employ pilot-scale testing prior to transition to commercial scale; they feel that they have a robust and well tested "scaled-down" model at the lab-scale (1–5 L). However, the only way to have a robust, well tested scaled-down process is to have scaled nearly identical processes in the past. Groups that advertise their ability to transition confidently, from 1 L bench-scale to commercial scale without an intermediate stage are inevitably scaling processes for producing a derivative product from a well-defined host and process (e.g., an antibody from CHO, or an enzyme from *E. coli*). As such, these fermentations are not being scaled for the first time, in reality, and many—if not all—of the scale parameters that will be discussed in this chapter have been defined during earlier iterations of the fermentation—probably via pilot-scale trials!

For companies developing truly innovative biobased products, using either "non-conventional" production hosts or employing novel, engineered biocatalysts, the knowledge base that fermentations producing "derivative" products enjoy simply does not exist and the behavior of the biocatalyst as scale increases cannot be predicted with any level of certainty. Transition to commercial scale directly from lab scale for these types of novel fermentation processes would be a risky endeavor.

6.5.1 Mixing

When scaling any fermentation process, the biggest challenge is culture mixing. This parameter is at the core of the majority of the challenges associated with the transition from lab scale to pilot and beyond (Junker 2004). At lab scale, achieving sufficiently good mixing that the culture broth can be assumed to be homogenous is taken for granted. As scale increases, though, mixing becomes a more difficult and complicated issue. Mixing time (which can be approximated by recording the pH change in a vessel when an addition of acid or base is introduced to the vessel in a location remote from the pH probe and determining the time required for the pH to stabilize at a new steady state) can comfortably be less than 1 s in a lab-scale process development fermentor but will increase to minutes at commercial scale (Table 6.1).

As mixing time increases, so too will culture heterogeneity; this heterogeneity is driven by both the biocatalyst utilization of nutrients from the medium and secretion of extracellular products. Cells remove nutrients—and, in the case of aerobic cultures, O_2—continuously from the medium; the rate at which this removal occurs being driven by the rate at which the cells are metabolizing (a factor related to maximum specific growth rate of the production organism and the physiological state of the culture). In small fermentation vessels, concentration gradients (in soluble nutrients or dissolved gases) are rapidly dispersed, due to the effective mixing. As fermentation scale increases and mixing becomes less efficient, significant concentration gradients develop, resulting in cells experiencing nutrient

Table 6.1 Representative mixing times and oxygen transfer rates in stirred tank fermentors

Scale	Volume (L)	Oxygen transfer (mmoles/L.h)	Mixing time (s)
Bench	<1–10	~400	~1
Pilot	1000–3000	~200	~30
Commercial	>50,000	~100	~100

limitation even though the "global" (or whole vessel) supply of nutrients and gases is sufficient.

The extent to which this issue is problematic is a function of both mixing time and utilization rates of the production organism. During periods of unrestricted growth, where all nutrients are available in excess, the maximum specific growth rate of the biocatalyst will dominate the rate of nutrient utilization. Maintaining nutrient supply to a rapidly growing organism (e.g., *E. coli* with a doubling time of <30 min) is more challenging than for a slow growing organism with a doubling time measured in hours. Aerobic prokaryotes are generally faster growing than eukaryotes and as such have higher requirements for mixing associated with O_2 supply. O_2 utilization rates in fast growing prokaryotic cultures can be very high (>300 mmol/Lh); although this O_2 utilization rate is often possible in lab-scale fermentors (Table 6.1), processes with such high metabolic rates are difficult to scale-up to commercial production due to both limitations in O_2 and heat transfer as scale increases (Junker 2004). Scale-up efforts (of aerobic fermentations) are most often hampered by the inability to achieve sufficient O_2 transfer as scale increases (Garcia-Ochoa and Gomez 2009). A large volume of work has been amassed associated with measuring and studying factors that impact the transfer of O_2 in fermentation tanks, and various methods for calculating the volumetric mass transfer coefficient (k_{La}) exist (Garcia-Ochoa and Gomez 2009). While optimizing the k_{La} of the fermentation is a worthy pursuit, a more pragmatic approach is to establish the maximum dissolved O_2 (DO_2) level that can be maintained with reasonable and scalable operational parameters (sparge rate, agitation, and back pressure) and how this DO_2 impacts process performance.

Nutrient and O_2 utilization is also related to the physiological status of the biocatalyst: if the growth is limited (by design) due to the exhaustion of a required nutrient—as is often the case when the target biobased product is a secondary metabolite—then the utilization rate of nutrients is usually decreased. This means that the nutrient and O_2 uptake rate (and therefore the need to establish efficient mixing) is at its maximum in a fermentation vessel just prior to the limiting nutrient depletion.

Due to the challenges presented by achieving efficient mixing, even with baffles and attempts to optimize agitation, commercial-scale tanks often suffer from the existence of stagnant zones, where the biocatalyst will experience nutrient depletion as nutrient utilization rates exceed the rate that these nutrients are replenished by medium mixing and diffusion. How a production host responds to intermittent nutrient depletion is a feature of strain robustness that is difficult to establish at lab scale, under well-mixed conditions, but will manifest itself as tank size increases. Conversely, if a fermentation control strategy calls for a nutrient limitation during a fed-batch process (e.g., a low residual glucose concentration), inefficient mixing can result in zones in the fermentor where the critical nutrient concentration is in excess of that required for the proper physiological state of the production strain.

As well as for compounds that are consumed by the production host, lack of efficient mixing also has consequences due to compounds secreted into the medium. These compounds can either be the desired product or a coproduct. They may exert direct toxicity if they accumulate (in a poorly mixed area of the vessel), or have indirect impacts, such as altering the culture pH. Again, bulk concentrations of compounds or pH (as determined by a pH probe at one or—preferably—several points in the fermentor) may be within acceptable ranges; however, in areas of a fermentation vessel that are less well mixed, these factors may be significantly outside the desirable ranges to a degree that they inhibit, or in extreme cases kill, the biocatalyst. It has been found that O_2 depletion (as a result of inefficient mixing) for as little as 15 s can impact cell metabolism.

Although increasing the mixing efficiency using a higher agitation rate seems to be a rational approach to overcoming longer mixing times as scale increases, this approach is limited in its applicability for both biological and economic reasons. In order to increase agitation, and therefore improve mixing, a higher power input is required which can render the fermentation process uneconomical. As fermentation vessel size increases, so do the diameter of the impellers. With larger impellers, the maximum shear and the variation in shear experienced at a given agitation rate increases. Maximum shear and shear variation are routinely higher at pilot and commercial scale than at laboratory scale, even at the decreased agitation rates used at larger fermentation volumes. Increased shear can directly impact the physiology and even viability of the production host (most notably, but not restricted to, filamentous organisms, *see below*). Excessive shear can also impact product quality, particularly for protein products and nutritional oils (the latter being released from the interior of the host, due to shear-induced cell damage, into the fermentation broth where chemical oxidation occurs).

Most scale-up approaches are centered on ensuring adequate fermentation broth mixing and gas transfer as scale increases. A wide variety of approaches have been suggested, tested, and applied to different production hosts and fermentation processes. Despite the successes of these different approaches for the reported examples, it is still true that no single, universally applicable approach to defining the best way to scale the mixing in fermentation processes has been devised (Garcia-Ochoa and Gomez 2009; Junker 2004). This is unsurprising, as the critical factors that need to be maintained constant (or as near to constant as possible) differ between both production hosts and desired target products. The challenge is, more often than not, determining which of the multiple factors (DO_2, nutrient supply, and shear) associated with mixing that is critical for the particular process being scaled and the best way to address this critical factor (or factors).

Mixing and nutrient gradients within a culture are problematic as scale increases for any fermentation process but, if the production host is filamentous, then the mixing issues can be significantly more involved. As well as complicating the efforts to obtain adequate mixing, filamentous production organisms are more prone to shear sensitivity than their unicellular counterparts (although unicellular organisms cannot be assumed to be shear-tolerant and this should be confirmed on a strain by strain basis). Filamentous production hosts are prone to incur cell disruption and impaired performance at tip speeds >3.2 m/s (Junker 2004). With fungal cultures, in particular, macroscopic morphology becomes a secondary factor that determines the broth viscosity and rheology. Morphology independently impacts the diffusion of nutrients and gases to and from living cells. Fungal cultures can grow in a variety of morphologies, the two extremes being: (i) completely hyphal, with the fungal hyphae not clumping and existing as individual filaments, or (ii) completely pelleted, where the hyphae are aggregated into discrete pellets ranging in size from less than a millimeter to several millimeters in diameter.

When non-pelleted morphology predominates, diffusion of nutrients from the media into the cells is not an issue. At low cell densities, this is the preferred form for hyphal cultures. As biomass density increases, however, ≥10 g/L dry cell weight, the hyphal cultures become viscous and difficult to mix, resembling oatmeal/porridge. As viscosity of the culture increases, maintaining a well-mixed culture becomes more difficult and the mass transfer decreases. Under conditions where a production host grows in a hyphal form, the biomass density that can be achieved, while maintaining effective culture mixing and homogeneity, is limited.

Hyphal cultures that adopt a pelleted morphology can grow to high cell densities (60–70 g/L dry cell weight) with the culture retaining a low viscosity and therefore being amenable to efficient mixing. Although the nutrient and gas composition of the bulk medium can be easily monitored and controlled under these conditions, the situation is not clear cut for the metabolically active biomass. Under conditions facilitating pelleted growth the culture becomes a heterogeneous culture consisting of discrete biomass pellets within a well-mixed and homogeneous bulk liquid medium. Diffusion of gases and nutrients across the pellet means that biomass in the interior of the pellet can experience nutrient depletion even while all nutrients are present in non-limiting concentrations in the culture broth. Pellets with a diameter greater than 1 mm can be expected to have interiors that experience nutrient depletion. As pellet size (and “compactness”) increase, so does the potential for cell starvation, due to nutrient gradients across the pellet. It is not unusual to observe “hollow”, necrotic interiors of larger more compact pellets.

Microbial macroscopic morphology is impacted both by the underlying physiology of the species being used and the culture conditions under which the organism is cultivated. There exists a large volume of the scientific literature dedicated to the factors (both media components and environmental factors) that impact morphology; review of this subject is well beyond the scope of this chapter. However, it is still the case that for most fermentations employing hyphal microorganisms, the best morphology (and the best way to achieve this morphology) is often identified empirically.

6.5.2 Gas Solubility

An unavoidable effect that is encountered as fermentation scale increases, and which can complicate process scale up, is the effect of hydrostatic pressure on the solubility of gases dissolved in the fermentation broth. This effect is most notable for CO_2 (when cultivating aerobic organisms) but could in principal occur for O_2. In a lab-scale fermentation vessel, there is no appreciable hydrostatic pressure, due to the height of the vessel (≤50 cm). Also, lab-scale vessels are commonly constructed of glass and not fitted with pressure rated seals, so back pressure is not routinely employed. However, at commercial scale, it is common practice to pressurize a fermentation vessel (≈5 psig, 1/3 atm.) in order to (i) decrease the chances of contamination issues by ensuring an efflux rather than influx of gas if any leaks occur; (ii) for foam mitigation; and (iii) to increase oxygen solubility to enhance oxygen transfer. Furthermore, a commercial fermentation vessel may be ≥30 feet (10 m) resulting in significant hydrostatic pressure. The use of back pressure and the height of the fermentation broth (hydrostatic head) mean that the medium at the base of the fermentor experiences a pressure in excess of one atmospheric pressure (2–4 atmospheres). Although in principle the increased pressure at the base of a commercial fermentation vessel could have a direct impact on the production host (this is rare at pressures <5 atm), it is far more common for a pressure effect to be due to an increase in dissolved CO_2. Elevated levels of CO_2 can negatively impact the growth and metabolism of some organisms; while this effect will not be evident at lab scale, during initial process development (or during strain development), the impact of dissolved gases on biocatalyst performance will start to be manifest at pilot scale, especially if (as is routine at MBI, *see below*) a satellite, lab-scale fermentation is carried out in conjunction with the pilot-scale test.

One of the present authors has personal experience of a process improvement program that was initially unsuccessful due to the impact of increased CO_2 in production-scale fermentors. An improved production strain was developed at lab scale, using classical selection techniques and validation of improved performance demonstrated in shake flasks and lab-scale fermentors. All performance targets were met by the newly developed strain; nevertheless, scale-up attempts were met with repeated failure. A detailed evaluation of potential reasons, including many of the other potential scale-up hurdles discussed here (shear, medium preparation, etc) finally identified that the improved strain was more sensitive to CO_2 than the incumbent production strain (confirmed using CO_2 shaking incubators). Following this epiphany, a subsequent strain improvement program, including a selection for strains tolerant to CO_2 at the levels anticipated at the base of the commercial-scale fermentation vessels, was pursued and improved strains were obtained that scaled up to pilot scale and beyond without significant issues.

6.5.3 Medium Preparation

It is usual for medium to be sterilized by autoclaving at laboratory scale, due to the limited volumes required and the modest size and portability of the fermentation vessels. Autoclave cycles in a lab-scale autoclave have a relatively short temperature ramp time (10–30 min) to 121 °C, a hold time of 20–30 min at this temperature, and a 20–30 min cool down time. Sterilizing a pilot or commercial-scale fermentor is carried out often using a "sterilize in place" (SIP) protocol. The medium is heated, either by direct injection of steam into the medium via the sparge or into the jacket of the fermentor. Medium additions and glucose feed solutions are either autoclaved (if the volume is sufficiently small) or sterilized separately in other feed tanks using a similar SIP protocol. The result of the difference in (i) volume and (ii) sterilization method between laboratory and pilot scale is an increase in the thermal exposure as scale increases due to both increased sterilization times and higher sterilization temperatures at large scale. Increased sterilization temperatures and/or times are employed as fermentation volume increases to ensure the entire medium is effectively sterilized. Increased exposure to thermal stress increases the potential of medium compound degradation for more temperature sensitive components. Often the most obvious consequence is the increased caramelization of glucose as the thermal exposure increases, resulting in a slight yellow-brown color for the sterilized glucose solution. This increased thermal degradation can be exacerbated by sterilization in metal tanks at pilot/commercial scale, rather than the more inert, predominantly glass vessels used at lab scale. Degradation of medium components can have a negative impact on fermentation performance in two ways: the destruction of nutritional compounds required by the production host, or generation of inhibitory degradation products (such as hydroxymethylfurfural, HMF, from glucose).

Although increasing the thermal exposure of culture medium can be simulated by increasing the cycle time in a lab-scale autoclave, this is not representative of the conditions experienced in a large-scale metal fermentation vessel and is only instructive at best. The true test is when medium is prepared for the first time in a pilot-scale fermentation vessel. If increased thermal exposure is identified as a major factor limiting scale-up, options to overcome this issue—such as using a continuous sterilizer—exist. Knowledge as to whether or not SIP or continuous sterilization is the most appropriate option for medium preparation can then be factored into either the design of a commercial facility or the selection of a toll facility.

6.5.4 *Product Separation and Purification*

Developing a fermentation procedure that achieves the required performance (in terms of desired chemical titer, productivity, and yield), using a medium that has a suitably low cost and that is scalable to pilot and commercial volumes, is only half the battle. The ability to produce a biobased chemical via fermentation does not mean that the process will be either commercially viable or scalable, as the final product is (usually) not the fermentation broth containing the product but a purified product meeting required product specifications. For that reason, in addition to a robust, scalable fermentation process, it is essential that a biobased chemical production technology also has a robust and scalable product recovery and purification process (often referred to as downstream processing or DSP).

The authors would strongly advocate that the potential issues and hurdles associated with product recovery and purification are addressed at an early stage of process development, as soon as a process for the production of the target chemical is being considered. The fact that a chemical can conveniently be produced via fermentation does not guarantee that it will be easy—or even possible—to recover in an effective and economical manner.

A conceptual route to product recovery should be in place before fermentation (or strain) development is initiated. Initial tests recovering commercially purchased chemicals (if available) from "test" fermentation medium (LB medium or similar) can be useful to convince researchers that the product can be recovered, even if purity and recovery yield optimization can wait until later in the development cycle. Once a fermentation process is sufficiently developed to be operated in a lab-scale fermentor, the work on the DSP can begin in earnest. This work should focus on product recovery yield and purity. The number of process steps should be restricted to the minimum necessary to achieve the desired purity as each unit operation adds costs and decreases recovery yield.

Although DSP technologies can and should be well developed before advancing a fermentation process to pilot scale, there remains significant advantages and benefits to the testing of the product recovery and purification at the pilot scale. At this scale, the unit operations used will be continuous (or semi-continuous) rather than the batch processes often tested at lab scale (e.g., for centrifugation) and will more closely resemble commercial scale operation and provide (high quality) equipment system design data. Furthermore, the energy requirements and mass balances are more conveniently and accurately calculated when operating at pilot rather than lab scale.

Finally, and often most importantly, pilot-scale demonstration of a process "soup to nuts" generates quantities of biobased product that can be tested by potential customers for "use compatibility". It is often the aim of biobased chemical technologies to provide a renewable alternative to a petrochemical-derived chemical feedstock. Confirmation that the biobased chemical can replace the petrochemical feedstock and that it can ultimately provide similar—and *ideally* identical—performance is crucial to obtaining customer acceptance of the biobased technology. Having a tub, jar, or bucket of the biobased chemical in its final form is also a powerful tool during investor meetings to give a sense of reality to the nascent technology. Although biobased chemicals may obtain the same chemical purity as the equivalent petrochemical source (e.g., >99.5 % chemical purity), this declaration of equivalent chemical purity is unlikely to convince potential customers of the biobased chemical's ability to replace the petrochemical equivalent. This skepticism is well founded as it is often the trace impurities found in the chemical (rather than the purity of the chemical itself) which determines the suitability of the biobased replacement chemical. In particular, organic sulfur (thiol containing amino acids) is often flagged as an issue for biobased monomers that undergo catalytic reduction or polymerization. Even trace amounts of these so-called "bad actors" can poison catalysts and make the chemical unfit for its intended use. The issues associated with organic sulfur-containing compounds in the fermentation broth that can contaminate the final product is one of the drivers (along with cost) for the removal of complex medium components and a desire for the development of simple chemically defined medium whenever possible. Often, only application testing is capable of establishing a chemical as "fit-for use", and these tests can require kilogram (or more) quantities of product—amounts that cannot be obtained using lab-scale fermentation and DSP equipment.

6.5.5 *Process Data for Further Scale-up Efforts*

An additional advantage of a well instrumented pilot-scale facility is that they provide a wealth of process data that can be used either to design a commercial fermentation plant or to define the requirements required in a toll manufacturer. Data generated at pilot scale (but not typically reliably captured at lab scale) include power consumption (for the agitator), chilling requirements, and suitable back pressure. The pilot-scale trials should (if well executed) generate a batch record and data package that will allow scale-up to commercial fermentors with a high degree of confidence.

6.6 Achieving Pilot Scale-up

6.6.1 *Use of Satellite Fermentations*

As is indicated above, there are a number of issues that can impact the performance of a fermentation process when it is transitioned from the lab bench to pilot scale. It is impossible to be sure ahead of time which factors may be critical in scaling up a biobased technology (if it were, there would be little or no value in pilot-scale trials). However, when scale effects are observed it is vital that efficient methods are in place to determine which of the multiple potential factors are the most important. It would require a large number of expensive experiments if all potential factors were tested in series at pilot scale. At MBI, a system of lab-scale satellite fermentations is routinely employed during initial pilot-scale testing to decouple potential sources of deviation from the (proven) lab-scale process and the pilot-scale trial performance (Fig. 6.1). Using parallel lab-scale satellite fermentors, it is possible to independently test seed train, medium preparation, and scale in a single experiment. Therefore, if the initial pilot trial is unsuccessful, it is easier to narrow down the potential issues underlying the difference in performance. If all the satellite fermentations achieve comparable performance to the lab-scale control, while the pilot scale trial does not, then the difference in performance can be confidently ascribed to a true scale-up issue (mixing, gas solubility, or shear). These can then be addressed in the subsequent pilot-scale batches.

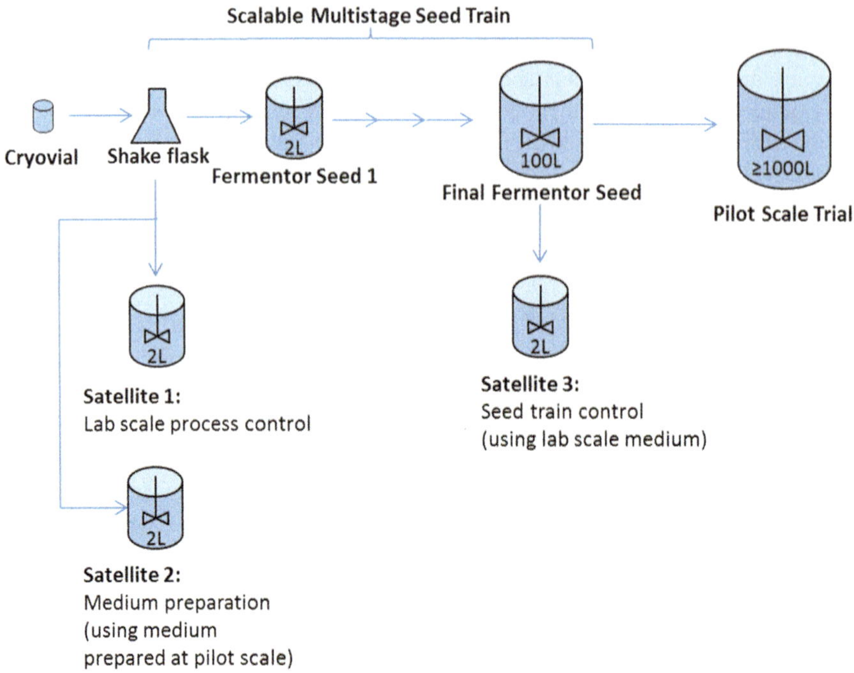

Fig. 6.1 The use of laboratory scale 'satellite' fermentations to decouple scale-up factors during transition of a fermentation from laboratory scale to pilot scale. *Satellite 1* acts as a lab-scale process control, confirming the performance of the cryo-vial stock and the initial shake flask seed inoculum; *Satellite 2* validates the medium preparation at pilot scale as it uses medium removed from the pilot scale vessel directly before inoculation; *Satellite 3* validates the multistage seed train and the quality of the inoculum used for the pilot scale test

6.6.2 Scale-up Strategies and Resources

Although a host of academic articles have proposed various approaches for the successful scale-up of microbial fermentation processes, from laboratory scale to pilot scale and beyond (Garcia-Ochoa and Gomez 2009; Junker 2004), it remains true that no single, robust, universal approach or algorithm for biobased technology scale-up has been developed. The scale-up demands of each technology are, if not completely unique, then are nearly so, and some degree of empirical experimentation (or just plain good fortune) will be required in the lab- to pilot-scale transition. Experience in scale-up is one way to help overcome some of this uncertainty; personnel that have scaled up multiple different processes can provide insights into the critical factors they have experienced in scaling technologies (similar if not exactly the same) as the biobased technology that requires scale up. Access to this kind of experience can be acquired by hiring in senior staff with the correct breath of experience. However, these people are often hard to find and expensive. A timelier and less expensive strategy (in the long run) is to access this kind of experience on an “as needed” basis, especially as this experience is only needed once for each technology scaled.

Fortunately for biotechnology companies looking to scale their biobased technologies, there are ways to access the required expertise. Consultants are a resource that can be used, if access to the equipment required for scale up can be sourced. Otherwise, there are a number of facilities in North America that provide as one of their services the scale-up of biobased technologies. These include the authors’ organization MBI (Lansing MI) as well as the University of Georgia, University of Maryland, Alberta Innovates, and ABPDU Berkeley (although the latter facility has a maximum fermentation vessel size of 300 L, which does not strictly fit into the definition of pilot scale used in this chapter and may be limited in its ability to replicate commercial scale). Organizations such as these have the advantage both in experience and expertise scaling multiple biobased technologies, and in the required equipment—fermentors from lab-scale to ≥1000 L (with the exception of ABPDU) to carry out the transition from lab to pilot scale. As an example, MBI has in the 5 years (between 2009 and 2015) transitioned 20 different fermentation processes directly from lab-scale to (3800 L) pilot scale without intermediate size. During this period, after addressing specific scale-related technical issues, all fermentations were successfully demonstrated at pilot-scale. These technologies scaled at MBI were for the manufacture of a wide variety of renewable products and employed a diversity of production hosts (fungi, bacteria, and yeast) as well as a wide variety of process control strategies (Fig. 6.2). Accessing this kind of external experience in scaling biobased technologies has advantages not only in maximizing the chances of success but also has advantages in terms of cost and timeline.

Fig. 6.2 The variety of processes scaled up to 3800 L in the MBI pilot between 2009 and 2014

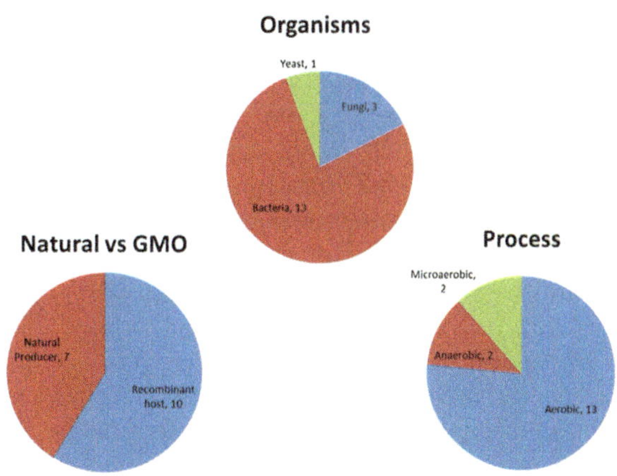

6.6.3 Financial Considerations

Pilot-scale demonstration of a biobased technology is a critical phase in the development cycle. This stage plays a vital role in demonstrating the robustness and scalability of the process, as well as provides kilogram quantities of material for application testing to demonstrate the utility of the biobased chemical. This transition to pilot scale involves many technical issues, but as described above, the approaches to identifying and overcoming these critical factors are available.

Along with the technical challenges associated with the transition from lab- to pilot-scale, there are financial and business challenges that must be overcome. In order to scale a process to a pilot fermentor, access to an appropriate pilot facility is necessary. Many companies initially plan to construct their own pilot plant (and many have done so); however, this can be a significant hurdle and may not be the most effective approach in terms of capital or time. To construct a pilot-scale facility will cost in the region of $10–$50 million (in the US), depending on the complexity of the process being scaled, and the design and construction could take 18 months at a minimum (after the capital has been secured). Convincing investors to part with the amount of money required can be problematic, based on lab-scale data alone, and even if this investment funding is forthcoming it could perhaps be spent more productively on further R&D. Building a pilot facility is not the end of the story. Experienced personnel will have to be hired to staff the facility, a process which again is time consuming and costly. Yet despite the challenges (funding, building, and staffing), the useful lifetime of the pilot facility is limited as (in an ideal word) the technology will rapidly progress past pilot to commercial scale. Once further scale-up (to commercial scale) is achieved, the pilot facility may be useful for testing further process improvements, but this activity is unlikely to make the pilot facility a sustainable asset. Few small- to mid-size biotechnology companies can expect to develop new technologies sufficiently frequently to justify the expense associated with the building and maintenance of a pilot facility.

The use of an existing facility—and an organization that specializes in biobased technology scale up—is a way to overcome (to some extent at least) the expense hurdle associated with technology demonstration at pilot scale. As these facilities are already built and operating, gaining access to this pilot-scale fermentation capacity has a far shorter timeline than construction of a new facility. Furthermore, as these fermentation plants offer their services to a host of clients (over many years), the cost at which access can be obtained is low in comparison to that involved in building a bespoke pilot facility. However, the most compelling advantage of using an external service provider for scaling up a novel biobased technology is that some of these facilities retain experienced and knowledgeable personnel who have scaled varied fermentation processes multiple times. It is thus the expertise, as well as the physical fermentation capacity, that is accessed. Using these facilities allows scale-up of fermentation processes to be achieved rapidly (3–

6 months), at lower cost (<$500,000), and negates the requirement for procuring investment to build a pilot plant facility and assembling a technology scale-up team that may have limited value once the process is scaled beyond pilot scale to commercial scale.

References

Garcia-Ochoa F, Gomez E (2009) Bioreactor scale-up and oxygen transfer rate in microbial processes: an overview. Biotechnol Adv 27:153–176

Junker BH (2004) Scale-up methodologies for *Escherichia coli* and yeast fermentation processes. J Biosci Bioeng 97:347–364

Index

S. Van Dien (ed.), *Metabolic Engineering for Bioprocess Commercialization*,
DOI 10.1007/978-3-319-41966-4

CPSIA information can be obtained
at www.ICGtesting.com
Printed in the USA
BVOW10*0217020816
457642BV00003B/27/P

9 783319 419640